Astrophysics and Space Science Proceedings

Volume 48

More information about this series at http://www.springer.com/series/7395

Andrea Orlando
Editor

The Light, The Stones and The Sacred

Proceedings of the XVth Italian Society of Archaeoastronomy Congress

Springer

Editor
Andrea Orlando
Catania Astrophysical Observatory
Catania
Italy

ISSN 1570-6591 ISSN 1570-6605 (electronic)
Astrophysics and Space Science Proceedings
ISBN 978-3-319-54486-1 ISBN 978-3-319-54487-8 (eBook)
DOI 10.1007/978-3-319-54487-8

Library of Congress Control Number: 2017937930

Printed on acid-free paper

This Springer imprint is published by Springer Nature
The registered company is Springer International Publishing AG
The registered company address is: Gewerbestrasse 11, 6330 Cham, Switzerland

Foreword

The main purpose of the XV meeting of the Italian Society for Archaeoastronomy (SIA) was the same as for the previous conferences, that is, to facilitate the communication between scientists and humanists as regards all that pertains to the cosmos. The meeting was held in the Monastero dei Benedettini, University of Catania, from 11 to 12 September, 2015. We thank heartily the Institutions of Catania for the help in the organization.

In the occasion of the meeting, we commemorated Prof. Gustavo Traversari, founder fellow of SIA, passed away in April 2015, and here, I would like to remember the importance of his figure. Traversari was full-time professor of Archaeology and of Greek and Roman Art History at Ca' Foscari University, Venice, from the beginning of '70s to 1999. His main researches dealt with the Greek and Roman sculpture and the Roman portrait, and he was interested also in the collecting in antiquity and modern times. He organized many scientific initiatives and founded the "Rivista di Archeologia" (Journal of Archaeology) and the series "Collezioni e Musei Archeologici del Veneto" (Collections and Archaeological Museums of Veneto), with the publication of antiques of many museums and institutions of that region. He founded and for many years was director of the "Dipartimento di scienze storico-archeologiche e orientalistiche" (Department of historical, archaeological and oriental sciences), and he realized the Library, very rich in books and specialized periodicals. He organized archaeological missions in Venice and abroad, and some of those initiatives were supported by the CeVeSCO (Venetian Centre of Studies and Researches in Oriental Civilizations) founded in 1985. Thanks to his activity, the Venetian archaeology was enriched in short time of disciplines such as Medieval Archaeology, Etruscology, Prehistoric Archaeology, Archaeology of Roman Provinces, and Minoan-Mycenaean Archaeology.

Traversari along with Giuliano Romano and other astronomers and archaeologists organized meetings dedicated to archaeoastronomy and cultural astronomy. An international colloquium on archaeology and astronomy was organized in Venice in 1989, and the proceedings were published in 1991 in the "Rivista di Archeologia". Three meetings were organized at the Lincean Academy in the second half of nineties, and they were followed by the foundation of SIA in the year 2000. Thanks to Traversari's generosity, the CeVeSCO funded also the publication of the Rivista Italiana di Archeoastronomia (Italian Journal of Archaeoastronomy). Unfortunately, Traversari, due to circumstances, was forced to stop his activity several years ago, and SIA could no more take advantage of his invaluable advices.

Milan, Italy

Elio Antonello
President of SIA

Preface

In recent years, archaeoastronomy, and more generally cultural astronomy, has had great visibility and consideration in Sicily, and this thanks to scientific and cultural activities triggered by the writer and the Institute of Sicilian Archaeoastronomy, a cultural association born in 2014 based in Novara di Sicilia, a small village located in the province of Messina.

The Fifteenth Congress of the Italian Society of Archaeoastronomy (SIA) is the first annual meeting which takes place in Sicily in the history of the SIA; however, it is in Sicily that several archaeoastronomy's pioneers began the studies of this multidisciplinary science. In the nineteenth century, the German archaeologist Heinrich Nissen began a campaign of studies on the orientation of temples of many Greek cities in Sicily, work that was published in his book *Das Templum* and in several articles published in the journal *Rheinisches Museum für Philologie*. Moreover, at the end of the nineteenth century, Koldewey, Puchstein and Penrose realized others archaeoastronomical studies on the Greek temples in Sicily.

After about 100 years, in the late twentieth century, there were new archaeoastronomy's studies in Sicily, works done by Profs. Sebastiano Tusa, archaeologist and current Superintendent of the Sea of the Sicilian Region, and Giorgia Foderà Serio, astronomer of the Palermo Astronomical Observatory now retired. The studies of Profs. Tusa and Foderà Serio were dedicated to the orientation of Sesi of Pantelleria and of Sicilian prehistoric tombs, mainly to the shaft and rock-cut tombs.

In 2009, during the International Year of Astronomy launched by the UN to celebrate the four hundredth anniversary of the first astronomical observations with instruments made by Galileo Galilei in 1609 and the first findings obtained with these observations, I had the idea of creating a winter festival of cultural astronomy that in 2016 celebrated its seventh consecutive year. This festival, which is held every year in different historical places being itinerant, has been enormously successful, and helped to disseminate to the public the ancient astronomy and archaeoastronomy. Moreover, in 2012, I created another festival dedicated to the 'science of the stones and the stars': the summer festival entitled 'Stones and Stars'.

This festival, which in 2016 has become biannual, has involved thousands of people.

Following these activities, the Institute of Sicilian Archaeoastronomy has enabled prestigious research collaborations with Sicilian organizations and foreign universities; in particular, I want to mention the collaborations concerning: the study about the orientation of the Greek temples of Agrigento, held together with the Politecnico of Milan and the University of Waikato, and the study of the so-called megalithic spiral of Balze Soprane, realized together with the Superintendent and the University of Catania.

I hope that after this meeting and future studies, the archaeoastronomy can earn a worthy space within one of the universities of Sicily.

Catania, Italy

Andrea Orlando
Editor and Chair of SIA2015 Meeting

Contents

Part II Cultural Astronomy

Part I
Archaeoastronomy in Sicily and in the Mediterranean

Archaeoastronomy in Sicilian Prehistory

Sebastiano Tusa

Abstract In this paper I present a report on archaeoastronomy's studies conducted in the late twentieth century on some Sicilian monuments of prehistoric times. Moreover I present some considerations on the Serraglio area in Pantelleria, a place that could have an interesting archaeoastronomical valence. The studies on the Sicilian prehistoric monuments have been conducted together with distinguished colleagues such Giorgia Foderà Serious and Michael Hoskin (Tusa et al. 1992; Foderà Serious and Tusa 2001). These studies have focused on the Copper Age rock-cut tombs (4th–3rd millennia BC), the pseudo-megalithic funerary epigeic structures (end of 3rd–beginning of 2nd millennia BC), the pseudo-dolmen (final III–beginning II millennia BC) and the Sesi of Pantelleria (first half of the second millennium BC) in Sicily. The Serraglio valley in Pantelleria is very attractive because on it were identified a system of alignments made up of several large boulders fixed vertically to the ground. These 'menhirs' will be protagonists of a future multidisciplinary study.

1 Introduction

The debate on the presence of intentional constructive alignments influenced by astral guidelines as part of the prehistoric monumentality in Sicily has recently livened up thanks to numerous proposals from both the astronomical specialists community and by amateur archaeologists. The official archaeological environment has often responded with skepticism if not with open opposition.

Personally, thanks to my innate curiosity, I always tried to avoid preconceived negationist position. On the contrary I tried to understand the arguments with respect in a climate of serene dialectic because in my opinion, it is essential and necessary between those who are animated by a real passion for history and the

S. Tusa (✉)
Soprintendenza del Mare, Assessorato per i BB. CC. AA. e I.S., Regione Siciliana, Sicily, Italy
e-mail: stusa@archeosicilia.it

A. Orlando (ed.), *The Light, The Stones and The Sacred*, Astrophysics and Space Science Proceedings 48,
DOI 10.1007/978-3-319-54487-8_1

study of past civilizations. Also I cannot forget that in the past I was fortunate to make systematic research in the field of archaeoastronomy in Sicily and Tunisia in collaboration with two exceptional masters such as Michael Hoskin and Giorgia Foderà Serio.

Thanks to that experience, as well as to the collected information, I acquired a method that I consider necessary to address such issues (Foderà Serio and Tusa 2001). On the base of such experience and knowledge I am often skeptical towards the "discoveries" of alignments in prehistoric monuments or emerging elements of the landscape. I have elsewhere argued my skepticism towards the supposed alignment proposed for the "pierced stone" of Jato Valley, shown locally as "u campanaru", as well as for the equally alleged "Minoan tholoi" at Gurfa of Alia and ultimately also for the "chair of the King", in the Ficuzza woods not far from Corleone (Tusa and Vassallo 2012).

My point of view is well known and is based on the absence in the above mentioned places of any specific contextually with archaeological sites beyond generic and sporadic discoveries of artifacts of different ages of which Sicily is fitted almost anywhere. The contextual archaeology is a goal that our discipline has painstakingly achieved after decades of research and excavations summarily conducted. Contextual archaeology means not only the proximity of levels, artifacts and structures linked by obvious stratigraphic relationships, but also the necessary territorial contextually based on obvious inter-relational cultural logics. In other words, for example, decades of research and studies on the interrelations between the Aegean world and Sicily have now provided a clear picture of the dynamic contact between the two areas of the Mediterranean along with its chronological coordinates as well as socio-anthropological features and material evidence. It is well known that such contact had a long history that we can divide in different phases.

During the early centuries of the second millennium BC we have archaeological indicators that show the presence of the Syrian-Palestinian, Cypriot and, of course, Aegean-Peloponnesian trade elements of Meso-Helladic tradition. The arrival on the scene of the Mycenaean in the centuries around the middle of the same millennium is clearly attested through a variety of sites scattered mainly in the centre and east of the island and the Aeolian archipelago. One of the most ancient sea route connecting Sicily with eastern Mediterranean involved cabotage along the North Africa coasts and the passage through Pantelleria. In a later phase a sea route coming from Greece, through Apulia and southern Italy, was reaching the eastern Sicily and the Aeolian archipelago (Marazzi and Tusa 2005).

This framework is now well corroborated by a large number of exhibits showing in a clear and unambiguous ways that there are many indicators of contact between the different regions of the Mediterranean. In this situation now well established, we know well how the funeral monumentality develops in contact areas between indigenous-Sikani (of Castelluccio culture tradition) and the Aegean and eastern newcomers. According to the data collected among the cemeteries scattered in the vast territories ranging from the coast of Syracuse to the Agrigento and Caltanissetta areas, you can see a well defined architectural funerary typology which has nothing in common with the alleged tholoi of Gurfa that, on the contrary, could be well framed in the typology of medieval barns of North African tradition introduced in Sicily after the Arab-Berber invasion over the island in the XIX century.

This is how, in my humble opinion, the thorny problem of identification of chronology, function and cultural affinity of any archaeological evidence must be faced. We must necessarily base our analysis on the evidence of a double contextually: the stratigraphic and the territorial. In other words we have to avoid the fascination of any easy theory based on weak scientific base.

However I disagree with a denial position of the possible presence of archaeoastronomical evidence in the Sicilian prehistory. Such negative position was, in my opinion, stimulated by the almost total absence in Sicily of megalithic prehistoric monuments (Whitehouse 1981a, b; Tusa 1999, 2009). It is well known, in fact, that there is a strong coincidence of astronomical archaeological evidence with megalithism (Daniel 1958). Given the almost total absence of such megalithic evidence in Sicily, it was difficult to enucleate examples of archaeoastronomy in the largest island in the Mediterranean.

But if a pure megalithic evidence is absent in Sicily, we have many examples of mixed architectures (hypogeic and epigeic) that we can classify in a category defined pseudo-megalithic, whose orientation could be determined by choices and not by environmental influences (Tusa 2014a, b). Moreover, even some hypogeic funerary architecture, for particular environmental conditions, could have a definable orientation by choice. Finally, recently after the exploration carried out in the island of Pantelleria we identified elements that, at the level of pure working assumption, could be analyzed and interpreted as possible intentional alignments.

According to the above mentioned methodological approach gained thank to the experience achieved working with the remembered two masters in archaeoastronomy—Michael Hoskin and Giorgia Foderà Serio—I have always been suspicious from the contagion of easy euphoria followed the identification of a star aligned to an artificial structure. There is always a star line with two points of any territory. Therefore, the approach must be diametrically opposed to the exercise to find a star line connected with an artificial structure. On the contrary we have to find inside a vast archaeological case study any clustering or groupings of orientation and, if there are, and are corroborated by statistically acceptable data, carry out an investigation to identify any astral coincidence. Obviously this is possible, in my opinion, only on monumental and structural emergencies whose chronological and cultural identification is certain and homogenous.

The methodological approach learned by the two above mentioned scholars and an almost thorough knowledge of Sicilian prehistory (Tusa 1999a) gave me the opportunity to highlight what I believe actually conditioned by the desire to align, or rather direct, in the creation of artificial structures, to points in space. Applying such methodology and knowledge I have identified some forms of hypogeic structures that, in my opinion, have an intentional orientation. In particular, my attention was focused on those few specimens of pseudo-dolmen made by a chamber cut into the rock with a dolmen corridor of access. In western Sicily such typology of funerary structure is related to the appearance of Bell Beaker Culture as well as in the previous Copper Age rock-cut tombs.

2 Copper Age Rock-Cut Tombs in Sicily (4th–3rd Millennium BC)

The type of funerary hypogeum made by a rock-cut tomb accessible from a cylindrical recess, which supplants the simple grave pit dug in the ground, appeared in Sicily in the fourth millennium BC. It is likely that the first form of this tomb consisted of a little cave, which was reached by a small portal practiced in a small vestibule, which generally opened on slightly sloping. To this type belong, for example, the tombs of Tranchina (Tinè 1960–61) (Fig. 1), near Sciacca, of Ribera (McConnell 1988) (Castle district) and of Partanna (Corso Vittorio Emanuele). The portal of the cell was directly cut on the wall of a small recess, which constitutes the prototype of the later cylindrical recess regularly cut on a horizontal rocky surface.

It is likely that the emergence of the rock-cut tomb that replaces in Sicily the grave simply dug into the soil has to be related to changed religious needs and rituals (Tusa 1991). In this new type of tomb the portal and closing stone slab of the real funerary cell perform the function of constant visual identification of the tomb

Fig. 1 Tranchina (Sciacca, Agrigento), the graveyard with rock-cut tombs and cylindrical recess (photo by Michele Termine)

Fig. 2 Roccazzo (Mazara del Vallo, Trapani), rock-cut tomb with cylindrical recess (T40) (photo by Michele Termine)

as a cohesive element for the clan or tribe. A lithic ballast was placed at the bottom of the cylindrical access to better protect the tomb from desecration avoiding an easy removal of stone slab used to close the door portal.

The first type of rock-cut Sicilian funerary hypogeic tomb is, therefore, characterized by two basically important factors for the object of this essay: the constant visibility of the tomb, and, therefore, also of the door, and the possibility to choose its orientation. Inside the cylindrical access, the burial chamber could be oriented towards a predetermined location and desired horizon without any conditioning.

In Roccazzo (Tusa 1988; Tusa and Di Salvo 1988–89) (Fig. 2) and Tranchina graveyards the orientation (from the inside) of the cells, completely free from constraints, is distributed almost always to "sun rising—sun climbing—sun culmination". It means between the sunrise at the summer solstice and the culminating at midday (South).

At Tranchina, among 28 measured tombs, 26 have azimuth between 68° and 182° (3 in the 1st quadrant, 22 in the 2nd). Only two tombs, very close to each other, differ from the distribution with azimuth of 239° and 249° respectively. At Roccazzo among 34 measured tombs, 27 are oriented towards "sun rising—sun climbing—culmination sun", being between 65° and 201°; 4 are more towards North of the direction of the sunrise at the summer solstice (about 60°) and 3 fall in the 3rd dial with azimuth 211°, 216° and 217°.

The similar distribution of the orientation of the tombs of Roccazzo and Tranchina is not accidental but deliberate and further reinforcing the closeness, if not the cultural identity, of the two sites.

The obtained values indicate that the orientation of the tombs of the two sites fits well in the typical pattern common in the Mediterranean, which often sees the prehistoric tombs oriented between the sunrise and its maximum heyday. In this regard we recall the cases of graves of Pranu Mutteddu in Sardinia (Hoskin and Zedda 1997), the gallery graves in the eastern Catalonia which, however, were partially contemporaneous with those of the two analyzed Sicilian sites (Hoskin and Palomo i Perez 1998), of the eastern Catalonia dolmens (1998) as well as those of the megalithic tombs (dolmen) of Rio Gor in Almeria (Hoskin et al. 1994). Comparisons concern convergence in the orientation, but not the cultural aspects; but they could indicate affinities or general homogeneity of religious rituals and liturgies in a framework of clear cultural polymorphism. But you can see a phenomenal convergence, because the evidence described are located at an early stage of the development of social aggregation with graves still rigidly occupied by a single body, except for very rare cases.

With the end of Copper Age, around the middle of the third millennium BC (Culture of Malpasso), multi chamber tomb spreads in Sicily (Albanese Procelli 1992). In the Conca d'Oro and nearby areas we find this type of tomb in the necropolis of Carini-Ciachea, Uditore and Via Roma, in the Palermo urban area (Cassano et al. 1975). These tombs are made up of cylindrical access onto which up to four cells, often arranged on different levels, open. Predictably the orientation arrangement of cells is extremely heterogeneous. In the sample analyzed (necropolis of Ciachea), in fact, 9 cells are oriented towards the 1st quadrant, 4 towards the 2°, 8 towards the 3° and 1 to the 4°. This shows that the insert of new religious ideologies that influenced the orientation of early Copper Age tombs had a short duration. Already at the end of the third millennium BC in Sicily the appearance of graves with multiple cells and the custom of carving the rock-cut tombs in the vertical cliffs show us that the previous orientation preference between the sunrise and its maximum apogee, with all the ritual consequences, has disappeared. New rituals emerge as result of migration and strong cultural influences from Aegean and Eastern Mediterranean that imply different liturgical forms and religious content.

3 Pseudo-megalithic Funerary Epigeic Structures (End of 3rd–Beginning of 2nd Millennium BC)

When the rock-cut tomb accessible from a cylindrical recess disappear, at the end of Copper Age (end of 3rd millennium BC) we are witnessing, especially in the south-western Sicily, the phenomenon of the presence of well-known Bell Beaker Culture (Tusa 1987, 1993). In this period and cultural environment a new type of tomb emerges. It is made by a long built dromos, which can be configured as a true

Fig. 3 Marcita (Castelvetrano, Trapani), rock-cut tomb with dolmen corridor (photo by author)

dolmen corridor, juxtaposed to the classical rock-cut tomb widely present since final Copper Age and, above all, throughout the entire Bronze Age. These are, in particular, the cases of the tombs with dolmen corridor of Pergole (Mannino 1971), Marcita (Tusa 1986, 1997) (Fig. 3), Torre Donzelle (Conte and Tusa 2012) and many other sites especially in the area of Partanna. The presence of the dolmen corridor leads us to assume that it was possible to select an orientation. It is clear that the consequent orientation intentionality is linked to the presence of Bell Beaker Culture that is now ascertained was coming to Sicily from Sardinia. In fact the same type of tomb was found in Sardinia in association with Bell Beaker Culture (Tanda 1984).

The orientation of these tombs is totally different than the earlier (rock-cut accessible from a cylindrical recess of Copper Age). These tombs with dolmen corridor have an orientation that will stagger between the second and third quadrant (between East and West).

It would be interesting to provide parallel with Sardinian or Spanish tombs or other regions of Europe likewise classifiable in the Bell Beaker Culture to understand if such orientation may actually be a distinctive feature of this culture. We found a coincidence only with the tombs in the eastern Catalonia (Hoskin and Palomo i Perez 1998). However, these comparisons should be made with caution since the chronology is completely different (the Catalan graves are much older being dated between 3300 and 2700 BC). The more archaic Catalan tombs might suggest that these are prototypes in the generation and dissemination of beliefs that may underlie this orientation choice. This hypothesis is supported by the statement that it is the Catalan arch to be one of the most interesting area of the dissemination of the Bell Beaker Culture as it is from here that branched off its penetration line to the Balearic Islands, Corsica, Sardinia and Sicily.

Returning to the tombs with built dolmen corridor of western Sicily we assume that the traditional Sicilian hypogeic attitude based on rock-cut tombs is by no means supplanted by the insertion of megalithic schemes. However, the insertion of the epigeic corridor dolmen, that could be considered as a megalithic ancestry, brought in Sicily also symbolic and spiritual values and characters that were reflected in the orientation choice (Nicoletti and Tusa 2102). In fact, it is obvious that the arrangement of these tombs was done in full respect of intentionality based on the presence of built corridor dolmen.

The inclusion of corrective megalithic element, such as the corridor dolmen, is not just a mere formal loan. This is proved by the fact that the tombs of the same area, chronology and culture (Bell Beaker/Naro-Partanna), albeit without built dromos, but with dromos carved into rock, standing in almost flat areas, are oriented in the same direction. The results of the measurements performed in the choices tombs for the survey, all located near Partanna, are presented in Table 1.

Table 1 Data obtained for the 11 tombs near Partanna

	Az (°)	H (°)	δ (°)	A* (°)
Torre Donzelle	185.0	0.5	−51.6	188.1
Pergole	115.0	5.0	+16.2	110.7
Stretto	102.0	3.0	−7.5	99.6
Stretto	139.0	1.0	−35.9	137.8
Corvo	220.0	1.0	−35.9	220.9
Corvo	225.0	0.5	−33.8	225.3
San Martino	142.0	0.5	−38.1	141.4
San Martino	170.0	0.0	−51.2	170.0
San Martino	115.0	3.0	−17.5	112.4
Cisternazza	191.0	6.0	−45.2	206.2
Tomba del Coniglio	330.0	2.0	−44.8	332.9

Legenda Az Azimuth, *H* altitude above the horizon, δ declination and *A** Azimuth corrected by the magnetic declination

4 Pseudo-dolmen (Final III–Beginning II Millennium BC)

It was discussed at length the problem connected to the limited number of megalithic monuments in the Sicilian prehistory. Frankly you cannot give a precise explanation to this phenomenon. It impresses the nearby irrepressible and gigantic megalithic Maltese monumentality against the "Sicilian megalithic silence". This opposition is more macroscopic and wide in a larger Mediterranean panorama that includes regions with considerable megalithic presences such as Sardinia and Apulia. You might think of an almost total destruction of Sicilian megalithic monuments due to the particular island population density and the intense agricultural transformation undergone over the millennia. However I am inclined more towards a real peculiarity of Sicilian civilization usually linked to the cult of the dead based on the concept of the return to the motherland rather than associated with the air element. The greater cultural permeability of Sicily to the Aegean-Eastern cultures had to play a significant role in comparison with a minor European influence as in the case of continental megalithism.

However, we cannot even consider for Sicily a functional view of the megalithic monuments as that given by Renfrew (1972) about Maltese evidence. Renfrew attributed to the Maltese megalithic temples the function of "parish temples" as the expression of distinct cantonal groups. According to this theory the "megalithic explosion" in Malta was stimulated by the need to transform the conflict between the cantons, which often in a confined environment like a small island can lead to cruel conflict, in constructive competitiveness by raising from the ground more and more impressive and gigantic structures. In other words the various cantonal groups that coexisted in the archipelago expressed their position of rivalry rivaling on the floor of the monumental dimensions (Renfrew 2007). This interpretive model allows us to justify the absence of megalithism in Sicily since the island's size and its limited population in prehistory did not require such a prophylactic antidote to avoid lacerating conflicts.

Very few, therefore, are the megalithic monuments in Sicily, and those few are beyond the canonical typological definition of megalithic structures. Therefore we prefer to call them pseudo-dolmens because their characteristics are not perfectly typologically classifiable in the category of "dolmen". I will cite a few examples that can be encompassed in the late third and early second millennia BC and they can be justified taking into account the likely flow of cultural elements (and perhaps ethnic) that made landing in Sicily the Bell Beaker Culture.

One of the most western of these pseudo-dolmen (originally built in the last centuries of the third millennium BC) is at San Giorgio, near Sciacca. It was carved from a large boulder hollowed so as to create a real funerary circular chamber preceded by a large slab with dimples libation (Tusa 2014a, b) (Fig. 4). This structure cannot be defined as a classical dolmen demonstrating a strong adaptation to the characteristics of the local hypogeism. The position of the boulder in which the burial cave was excavated influenced the manufacture and, therefore, the opening side, which is directed, however, toward the sunrise.

Fig. 4 San Giorgio (Sciacca, Agrigento), pseudo-dolmen (photo by author)

In eastern Sicily we have the problematic evidence of Cava dei Servi, where a round dolmen shows a clear coincidence with the perimeter of rock-cut tombs. It consists of a circular regular arrangement of orthostats covered by one or more horizontal lintels (Di Stefano 1979; Belluardo and Ciavorella 1999) (Fig. 5). Another one, which has similar features, has been located at short distance from the aforementioned circular dolmens (Belluardo and Ciavorella 1999). The presence of Castelluccio pottery, albeit in a not perfectly predictable context, lead us to date such structures to the early second millennium BC.

A third example of pseudo-dolmen is present at Mura Pregne, near Termini Imerese, on the Tyrrhenian coast of the island (Fig. 6). It is a clear corridor dolmen that was violated long ago. It took advantage of the existence of a large existing natural boulder. The site presents evidence from prehistoric times up to the middle age; therefore its precise dating seems impossible (Ghizolfi 1993).

It is interesting to note that the two dolmens of Cava dei Servi are oriented substantially toward the sunrise at the summer solstice as the great mass of the European dolmen (Table 2). The evidence is so narrow that every deduction is improper, however it is interesting to note that while one of the two tombs is arranged in a "natural way", it means that is perpendicular to the slope of the hill in order to take a static advantage, the second is built in a different manner and arranged with the axis tangent to the wall proving an intentional orientation.

Fig. 5 Cava dei Servi (Rosolini, Siracusa), circular dolmen (photo by author)

Fig. 6 Mura Pregne (Sciara, Palermo), dolmen (photo by author)

Table 2 Data obtained for the 2 dolmens of Cava dei Servi

	Az (°)	H (°)	δ (°)	A* (°)
Cava dei Servi	64.0	4.0	22.9	60.6
Cava dei Servi	59.0	2.0	25.4	57.2

Legenda Az Azimuth, *H* altitude above the horizon, *δ* declination and *A** Azimuth corrected by the magnetic declination

5 The Alignments of Serraglio (Pantelleria) and the "Sesi"

In the southeastern portion of the island of Pantelleria, in one of the most picturesque areas where still intact is the traditional system of agricultural terracing with rural annexes ("dammusi" and "sarduni"), we identified a system of alignments made up of several large boulders fixed vertically to the ground (Fig. 7). The area is greatly important in the general economy of Pantelleria topography because it constitutes the only valley that is continuously connecting the two western and eastern coasts slightly to the south of the island midline. The Serraglio valley, where there are the alignments, is entering seamlessly into that vast plain of Ghirlanda that was and is the main resource for the island's agriculture.

The Serraglio Valley had a significant role in the early history of Pantelleria because it is here that since prehistoric times (mid-second millennium BC) the islanders stocked up on clay for the manufacture of ceramics. From here came the clay used during the late Roman period to produce large quantities of fire pottery exported throughout the central Mediterranean, defined in the archaeological literature "Pantellerian Ware". The clay is found abundantly in banks where it appears in association with secondary volcanic phenomena. This clay is just the product of sublimation phenomena generated by the leakage of vapors from multiple cracks in the ground.

Fig. 7 Serraglio (Pantelleria, Trapani), pseudo menhir (photo by author)

Fig. 8 "Favare" (Pantelleria, Trapani), stone circles (photo by author)

The presence of extensive secondary volcanic phenomena constitutes one of the most important attractions of this valley. In an elevated position on the right side of the valley there is a site named "Le Favare", where there is the presence of some fumarole with spillage of hot vapors. It is clear that the presence of this kind of phenomena may have captured the man's interest in the long history of Pantelleria giving the area a symbolic character.

However, apart from pottery of various ages (since prehistory to Roman period) collected on surface, extensive archaeological research was never carried out in order to understand the nature and function of the settlement. In the area of Favare there is superficial evidence of apparent alignments and stone circles that could indicate what remains of ancient structures unequivocally linked to the adjacent volcanic phenomena (Fig. 8). However it is clear that without any contextual and stratigraphic feedback it is impossible to hazard any conclusions concerning such structures that could be also the remains of recent land arrangements.

The area where we have highlighted the presence of some elongated large stones stuck to the ground and with apparent alignments is located in the highest of its downstream part below the area of Favare, shortly before the watershed between the valley of Serraglio and that of Girlanda.

On a slightly downward terraced slope from north to south, we noticed the presence of 14 stones of varying size (between 3 and 1.5 m tall with diameters oscillating around 1 m) whose definition as menhir is risky if by this term we mean

Fig. 9 Serraglio (Pantelleria, Trapani), pseudo anthropomorphic menhir (photo by author)

orthostats whose original appearance has been visibly altered by human hands, which generally has regularized its form (Fig. 9). In truth all of Serraglio elements have the peculiarity of having the distal tapered upper part, but all are, also, characterized by the absence of any conspicuous trace of modification by man. In other words we have elongated stones placed in an upright position, but that could also be entirely natural in their current texture and appearance. In truth there are two larger and closer menhir (4 m), presenting an eccentric tapering top that recalls the Sardinian anthropomorphic menhirs, where the lateral taper has been identified with the desire to represent the human figure.

The distances between the orthostats of Serraglio, apart from the two already described that are located at the beginning of one of the two alignments, are sometimes significant, but often are oscillating between 10 and 30 m. The two alignments, formed respectively by 4 (the West) and 8 (the eastern one) menhir, converge towards North. Beyond the point of convergence there are two other close orthostats that point towards the same direction.

We are, therefore, at the presence of two north converged alignments with a northern appendage similarly oriented, which is a sort of large arrow pointing north towards the Favare area (Fig. 10). And it is interesting to note a peculiar assimilation of this orientation with the famous menhirs alignment of Palaggiu in Corsica.

We started by making accurate surface surveys, despite the dense existing vegetation, and a stratigraphic sounding at one of two nearby orthostats (Fig. 11).

Fig. 10 Serraglio (Pantelleria, Trapani), aerial photo showing the pseudo menhir alignments (photo by author)

In both cases the result was positive but strange at the same time. We would have expected, in fact, to highlight elements that could bring us to the Early or Middle Bronze Age, the period of the most significant traces of prehistoric presence on the island. On the contrary either the sounding pit or the survey revealed the presence of a widespread occupation of the Roman imperial period. It is interesting to remember that near the alignment there are some ancient cisterns of the same period and a large stone with superior artificial cavities with obvious function of liquid container that could give rise to think of being in front of a ritual basin (Fig. 12). The above Roman traces could be classified as the usual rural settlement type. This evidence fits very well with settlement pattern of the first centuries of Roman Empire, scattered all over the island showing a strong population increase.

But the contextuality of standing stones with Roman era would suggest the presence of cults related to volcanic emergencies, albeit minor, or simple functional geographic indication to facilitate the path towards the Favare. The limited available data leads us to be cautious to interpret the alignment as a result of ritual activity connected with volcanic phenomena, although the Mediterranean is full of rituals emergencies related to such phenomena. In Pantelleria we have the sanctuary at the Lake of Venus that shows the presence of cults linked to local volcanic emergencies since the late third millennium BC up to Roman period (Tusa and Ursini 2012).

Among Pantelleria antiquities we cannot forget the presence of those unique megalithic monuments dated to the first half of the second millennium BC: the Sesi

Fig. 11 Serraglio (Pantelleria, Trapani), sounding near a pseudo menhir (photo by author)

(Orsi 1899; Tusa 2014a, b). Sesi are similar to the Sardinian Nuraghe (Contu 1981), to the Balearic Talayots only for a vague exterior formal shape (Fig. 13). The Sesi, in fact, do not present a hollow structure inside because they are made by a well-constructed veneer made by powerful blocks that encloses an internal huge mass of stones. One or more small cavities for funerary ritual are opened on its sides.

The Sesi arises from the need to transfer the type of Sicilian rock-cut tomb into the rocky environment of Pantelleria where the particular hardness of the volcanic rock did not allow the carving of the burial chambers with those instruments used in that period. The insertion of this peculiar type of funerary tumulus in Pantelleria represents the will of a population migrated from Sicily to transfer to the island an otherwise impractical funeral ritual: that of the rock-cut tomb carved into the rock.

Fig. 12 Serraglio (Pantelleria, Trapani), large stone with superior artificial cavities probably used as a ritual basin (photo by author)

Fig. 13 Mursia (Pantelleria, Trapani), Sese (photo by author)

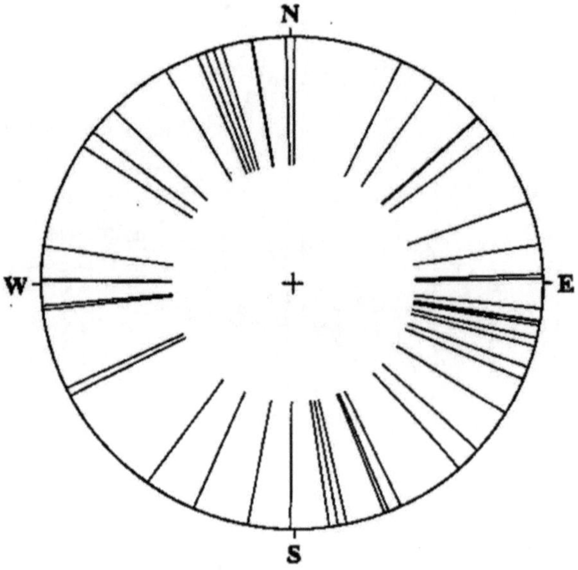

Fig. 14 Distribution of the orientations of the Sesi of Mursia (from Tusa et al. 1992)

According to these considerations we think that there is nothing of real megalithic features at the base of Sesi appearance, if we give to this term also a cultural and religious significance. It was predictable that this "formal and not substantial megalithism" had no need to orient its facilities to precise points of space. An analysis of 42 sesi, from which we have to exclude 18 due to bad conservation, doesn't allow any sensible measure; 49 measured cell' orientations show a clear absolutely heterogeneous distribution (Tusa et al. 1992) (Fig. 14).

6 Conclusions

Summarizing the analysis conducted on the possible indicative elements of a choice in the architectural orientation in the Sicilian prehistoric times, we must admit that the final result is not comforting. Apart from a few examples confined to certain times and to certain areas of the island, we can say that the phenomenon is in Sicily absolutely marginal.

It is clear that there is a link between the marginality of monumental orientation in Sicilian prehistory and similar marginality of megalithism. The convergence of these marginality between the megalithism and intentional orientation is particularly important because it confirms the idea that sees in megalithism not only a constructive method, but also the vehicle of religious ideologies that have particularly attention to the astronomical orientation of religious and funerary structures (Daniel 1941, 1967; Renfrew 1981).

Limiting ourselves, then, to those monuments whose features appear certain, we can say that both the megalithism and astronomical orientation are in the long and complex dynamics of Sicilian prehistory marginal and limited to short periods with restricted locations.

Both the scholar and the enthusiastic lover of the history of our country are disappointed that the fascinating field of archaeoastronomy is so marginal dealing with Sicilian prehistory. The method used does not allow me to go beyond what is succinctly expressed in this essay. But we hope that future discoveries could change this situation.

References

Albanese Procelli, R.M.: La necropoli preistorica di contrada Malpasso. Not. Sc. **XLII–XLIII**, 168–225 (1992)

Belluardo, S., Ciavorella, G.: Alla ricerca della grande Dea. L'architettura preistorica rupestre negli Iblei. Tra arte e religione. L'Erma, Roma (1999)

Cassano, S., Manfredini, A., Quojani, F.: Recenti ricerche nelle necropoli eneolitiche della Conca d'Oro. Origini **IX**, 153–271 (1975)

Conte, L., Tusa, S.: Resoconto delle attività di ricerca e scavo nel territorio di Partanna, TP (anni 2000–2005). Indagini nelle necropoli preistoriche di Stretto, Torre Donzelle e Capo d'Acqua, in Atti della XLI Riunione Scientifica dell'Istituto Italiano di Preistoria e Protostoria, Dai Ciclopi agli ecisti. Società e territorio nella Sicilia preistorica e protostorica (San Cipirello (PA), 16–19 novembre 2006), Firenze, pp. 719–727 (2012)

Contu, E.: L'architettura nuragica, pp. 45–62. Ichnussa, Milano (1981)

Daniel, G.E.: The dual nature of the megalithic colonization of Europe. Proc. Preh. Soc. **VII**, 1–49 (1941)

Daniel, G.E.: The Megalith Builders of Western Europe, London (1958)

Daniel, G.E.: Northmen and Southmen. Antiquity **XLI**, 313–317 (1967)

Di Stefano G.: La collezione preistorica della "Grotta Lazzaro" nekl Museo Civico di Modica. Sicilia Archeologica **41**, 91–110 (1979)

Foderà Serio, G., Tusa, S.: Rapporti tra morfologia ed orientamento nelle architetture rituali siciliane dal IV millennio al II millennio a.C., in Atti del 3° Convegno internazionale di Archeologia e Astronomia, L'uomo antico e il cosmo, Roma, Accademia nazionale dei Lincei, pp. 297–324 (2001)

Ghizolfi, P.: Mura Pregne. BTCGI **XII**, 129–138 (1993)

Hoskin, M.: Studies in iberian archaeoastronomy: (5) orientations of the megalithic tombs of northern and western Iberia. JHA **XXIX**, S39-S87 (1998)

Hoskin, M., Palomo i Perez, T.: Studies in iberian archaeoastronomy: (4) the orientation of megalithic tombs of eastern Catalunya. JHA **XXIX**, 63–79 (1998)

Hoskin M, Allan E., Gralewski R.: Studies in Iberian archaeoastronomy: (1) orientation of the megalithic sepulchres of Almeria, Granada and Malaga. JHA, **XXV**, S55–S82 (1994)

Hoskin, M., Zedda, M.: Orientations of Sardinian Dolmens. Archaeoastronomy (supplement to JHA) **22**, 1–16 (1997)

Mannino, G.: La tomba di contrada Pergola. Sicilia Archeologica **15**, 52–56 (1971)

Marazzi, M., Tusa, S.: Tokens, counters e altri dispositivi mnemotecnica fra Vicino Oriente e Mediterraneo nel II millennio a.C.: qualche riflessione alla luce dei nuovi ritrovamenti da Pantelleria, Studi egei e vicinorientali 1, Studi in onore di Enrica Fiandra, Napoli, pp. 163–190 (2005)

McConnell B.: La necropoli di contrada Castello (Ribera, Agrigento): lo sviluppo della tomba a grotticella con pozzetto. Rassegna di Archeologia **7**, 548–549 (1988)

Nicoletti, F., Tusa, S.: L'età del Bronzo nella Sicilia occidentale, in Atti della XLI Riunione Scientifica dell'Istituto Italiano di Preistoria e Protostoria, Dai Ciclopi agli ecisti. Società e territorio nella Sicilia preistorica e protostorica (San Cipirello (PA), 16–19 novembre 2006), Firenze, pp. 105–130 (2012)

Orsi, P.: Pantelleria. Monumenti Antichi dei Lincei **IX** (1899)

Renfrew, C.: Malta and the calibrated radicarbon chronology. Antiquity **46**, 141–146 (1972)

Renfrew, C.: The Megalith Builders of Western Europe. In: Renfrew, C. (ed.) The Megalithic Monuments of Western Europe, London, pp. 8–17 (1981)

Renfrew, C.: Ritual and Cult in Malta and beyond: traditions of interpretation. In: Barrowclough, D., Malone, C. (eds.) Cult in Context: Reconsidering Ritual in Archaeology, pp. 8–13. Oxbow Books, Oxford (2007)

Tanda, G.: Benetutti Loc. Maone. In: Anati, E. (ed.) I Sardi: la Sardegna dal paleolitico all'età romana, pp. 286–287. Jaca Book, Milano (1984)

Tusa, T.: Dinamiche storiche nel territorio selinuntino nel II millennio alla luce delle recenti ricerche in contrada Marcita (Castelvetrano). In: Marazzi, M., Tusa, S., Vagnetti, L. (eds.) Traffici micenei nel Mediterraneo, Palermo, pp. 133–140 (1986)

Tusa, S.: The Bell Beaker in Sicily (Italy). In: Waldren, W.H., Kennard, R.C. (eds.) Bell Beakers of the Western Mediterranean. BAR Int. Ser. **331**, 523–550 (1987)

Tusa, S.: L'insediamento di Roccazzo (Mazara del Vallo, Trapani): nuovi elementi di inquadramento ed interpretazione dell'Eneolitico siciliano. Rassegna di Archeologia **7**, 553–554 (1988)

Tusa, S.: Considerazioni sulla religiosità delle popolazioni pre-elleniche siciliane tra il Paleolitico superiore e l'età del Bronzo. ΜΥΘΟΣ **3**, 167–188 (1991)

Tusa, S.: Il Bicchiere Campaniforme in Sicilia: evento, congiuntura o dinamica strutturale. Studi sulla Sicilia occidentale in onore di Vincenzo Tusa, Padova, pp. 203–214 (1993)

Tusa, S.: L'insediamento dell'età del bronzo con Bicchiere Campaniforme di Marcita. Corrao editore, Trapani (1997)

Tusa, S.: The megalith builders and Sicily. J. Mediterr. Stud. **1**(2), 267–285 (1999a)

Tusa, S.: La Sicilia nella preistoria. Sellerio editore, Palermo (1999b)

Tusa, S.: Il megalitismo e la Sicilia. In: Tusa, S. (ed.) Le orme dei giganti, Palermo, pp. 21–24 (2009)

Tusa, S.: Apporti megalitici nelle architetture funerarie e abitative della preistoria siciliana. In: Gullì, D. (ed.) From Cave to Dolmen, Ritual and symbolic aspects in the prehistory between Sciacca, Sicily and the central Mediterranean, pp. 237–246. Archaeopress Archaeology, Oxford (2014a)

Tusa, S.: L'insediamento protostorico di Mursia (isola di Pantelleria). In: Tolve, A., Tusa, S. (eds.) Archeologia dell'insediamento protostorico di Mursia (Pantelleria, Italia), BAR Int. Ser. 2621, pp. 1–22 (2014b)

Tusa, S., Di Salvo, R.: Dinamiche funzionali ed organizzazione territoriale dell'insediamento eneolitico in Sicilia: l'evidenza di Roccazzo (Mazara del Vallo, Trapani). Origini **XIV**, 101–129 (1988–89)

Tusa, S., Ursini, D.: Rinvenimenti eneolitici a Pantelleria – Lago di Venere, in Atti della XLI Riunione Scientifica dell'Istituto Italiano di Preistoria e Protostoria, Dai Ciclopi agli ecisti. Società e territorio nella Sicilia preistorica e protostorica (San Cipirello (PA), 16–19 novembre 2006), Firenze, pp. 623–636 (2012)

Tusa, S., Vassallo, S.: Ma no! E' un monumento di età romana. Archeologia Viva XXXI **156**, 54–55 (2012)

Tusa, S., Foderà Serio, G., Hoskin, M.: Orientations of the sesi of Pantelleria. JHA **17**, S15–S20 (1992)

Whitehouse, R.: Megaliths of the Central Mediterranean. In: Renfrew, C. (ed.) The Megalithic Monuments of Western Europe, London, pp. 42–63 (1981a)

Whitehouse, R.: The rock-cut tombs of the central Mediterranean. Antiquity **XLVI**, 275–281 (1981b)

The "Campanari": Big Artificially Pierced and Astronomically Oriented Rocks in the South Territory of Monte Iato (Sicily)

Ferdinando Maurici, Vito F. Polcaro and Alberto Scuderi

Abstract In Western Sicily, in the province of Palermo, within the municipal borders of the towns of San Cipirello and Monreale, there are two large rocks with artificial holes, bot locally named "Campanaru" (i.e. "Bell Tower") at less than 8 km away from each other. One of these rocks is still existing and visible from far away. A lightning destroyed the second one in 1968 or a few years later. Both holes are undoubtedly artificial and astronomically oriented with extreme accuracy. The rock still existing on Monte Arcivocalotto has its hole aligned with the sunrise of the winter solstice, while the collapsed one, documented by a photo, by oral testimonies and existing remains, is sited on the Cozzo Perciata hill and had its hole axis exactly oriented at the sunrise of the summer solstice. For the latter perforated rock still exists the tradition that put in relation the sunrise into the hole at the summer solstice with the start of harvesting works which traditionally begin on a date close to the end of June. An Eneolithic/Early Bronze Age settlement is archaeologically well attested at the Monte Arcivocalotto site. Both from Cozzo Perciata and from Monte Arcivocalotto, Pizzo Pietralunga is clearly visible. This is an outstanding, insulated geological structure, on whose base is an Eneolithic/Early Bronze Age settlement, whose materials would indicate a cultic character attendance and/or an exchange site of the local populations. Also in close proximity of the Cozzo Perciata perforated rock are abundant fragments of pottery dating back to the Eneolithic and the Early Bronze Age, while a pseudo-tholos tomb with dromos was found at a few hundred meters. The fact that in this area there are two coeval and similar monuments (artificially perforated rocks) with different and complementary (winter and summer) solstice alignments seems to indicate that here, between the Eneolithic and the Early Bronze Age, a civilization has developed that had a solar calendar and developed a simple but very effective technology to

F. Maurici
Regional Cultural Heritage Superintendence of Sicily, Sicily, Italy

V.F. Polcaro (✉)
INAF/IAPS, Rome and ACHe, Ferrara University, Ferrara, Italy
e-mail: vitofrancesco.polcaro@iaps.inaf.it

A. Scuderi
National Vice-Director, Archaeological Groups of Italy, Rome, Italy

A. Orlando (ed.), *The Light, The Stones and The Sacred*,
Astrophysics and Space Science Proceedings 48,
DOI 10.1007/978-3-319-54487-8_2

materialize it. Other megalithic monuments, dating to the same period and showing astonishing hierophanies at the solstices sunrise, as the so-called "King's Pulpit" in the nearby park of the "Ficuzza Forest" and others, were found in the same geographic area.

1 Introduction

The *Rollo* of the Archbishopal Church of Monreale[1] is a document drawn in 1182, showing the boundaries and the description of the lands of the immense area of western Sicily assigned in feud and church dioceses to the Monreale archbishopric by William II. In particular, in the description of the territory identified as *divisa terrarum hospitalis Sancte Agnes,*[2] the existence of a *petra perforata*, part of a group of four *petras plantatas ordinatim posita*, is remembered. This is not the only case of special stones mentioned as reference points.[3]

Isolated stones, sometimes artificially stuck in the ground, and possibly artificial alignments of stones are then registered in the territory of Monreale in the twelfth century: in particular, the presence of a perforated stone caught the attention of the drafters of that famous document. Obviously stones *plantate* (=stuck) in some cases could also be natural, or they could be been stuck in the ground even in very recent times, as a signal for border demarcation. However, in most cases they had to be stones of a certain size, well visible and conspicuous in the territory. Every other consideration concerning the *petra perforata* in the *divisa terrarum hospitalis Sancte Agnes* would be arbitrary. We must thus limit ourselves to register the fact that at least a perforated stone appeared visible in this territory, in AD 1182.

However, a big rock with large artificial circular hole is still visible on Mount Arcivocalotto[4] (37° 55′ 10.34″ N–13° 14′ 15.46″ E), a few km from the southern slopes of Monte Jato, in Monreale municipal area, about 30 km south Palermo (Fig. 1). The site is not far from the medieval *divisa hospitalis Sancte Agnes* remembered in 1182, if the most accepted identification of the latter with the Sant'Agata district in the municipality of Piana degli Albanesi is correct.

---

[1]A digital version of the *Rollo* manuscript is present on the web at: http://vatlat3880.altervista.org.

[2]The term *divisa* identify a territory. The *divisa* of Hospital of St. Agnes is usually identified with the present day "contrada S. Agata" (IGM F° 258 I N.O. Piana degli Albanesi).

[3]In the *divisa Haiarzeneti*, not far from the *hospitalis Sancte Agnes*, the same document cites the presence of a *petram magnam plantatam*. Other *petras plantatas* are cited in the *magna divisa Iati*, the lands dependent on the fortified hill center of Iato (in arabic, *qal'a*). *Petre magne plantate insimul* in the *divisa Hendulcini*, and *petre plantate* in the *divisa* of *casale Benbark* and in other ones were also cited in the *Rollo.*

[4]The current name is easily connected to the ancient Arab names *Hagiarbukal* today Arcivocale district, and Rahalbukal, mentioned in the *Rollo*, near which stands Monte Arcivocalotto, actually a mere 569 m high hill. The Arcivocale town name comes from the Arabic *al-Hagar būqāl*, "Stone Mug" (Caracausi 1994, p. 68).

Fig. 1 The "Campanaru" of Monte Arcivocalotto

2 The "Campanaru" of Monte Arcivocalotto

This remarkable monument has been repeatedly published (e.g. Polcaro et al. 2012; Scuderi et al. 2013, 2015) and will therefore only briefly described. It is situated in the area of Sicani Mountains in Northwestern Sicily, in the part in which the Jato river flows into the Belice river in its northern branch, forming a line of communication between the interior and the southern coast of Sicily. The area, inhabited at least since Neolithic times, had a great development in the Bronze Age.

The site at the foot of Pizzo Pietralunga (Fig. 2), an isolated natural rock about 150 m high, which stands on the Belice Valley (Scuderi et al. 1997), is of particular importance.

In the Middle Ages this enormous isolated rock probably was an easy and immediate reference to the indication of the boundaries of the surrounding lands, all belonging to the Church of Monreale[5] (Nania 1995, p. 107). Considering the size, position and the certainly out of the ordinary profile of Pizzo Pietralunga and its visibility from long distance, it is easy to conclude that this extraordinary rock has most probably attracted the attention of the prehistoric inhabitants of the area.

At its feet it is in fact sited a wide archaeological area (4500 m^2) with Eneolithic Age and the Bronze Age materials (e.g. Scuderi et al. 1996, p. 21). The settlement

[5]In the famous *Rollo* of the Monreale Church the present *Petralonga* (Sicilian name for Pietralunga) is probably recorded as *Hagiarbukal* (see note 4).

Fig. 2 Pizzo Pietralunga

of Pizzo Pietralunga, would seem to have the site of Monte Arcivocalotto as hegemonic center, while the fineness and the particularity of the materials present at the feet of Pizzo Pietralunga, including ceramic of the Bell Beaker Culture very rare in this area, could indicate a particular role and function (religious and/or exchange area) acting within the district.

Not far away, the Monte Arcivocalotto site is placed in a dominant position, on the isolated hill at about 500 m above sea level. This hill is located at almost equal distance between the mountains of Rocca Busambra and Monte Jato and about 2.5 km from Pizzo Pietralunga. This settlement is known since some time, thanks to the surface presence of fragments dating from Eneolithic to the Bronze Age. The site was later occupied in the Roman, Byzantine and Medieval period (Scuderi et al. 2011).

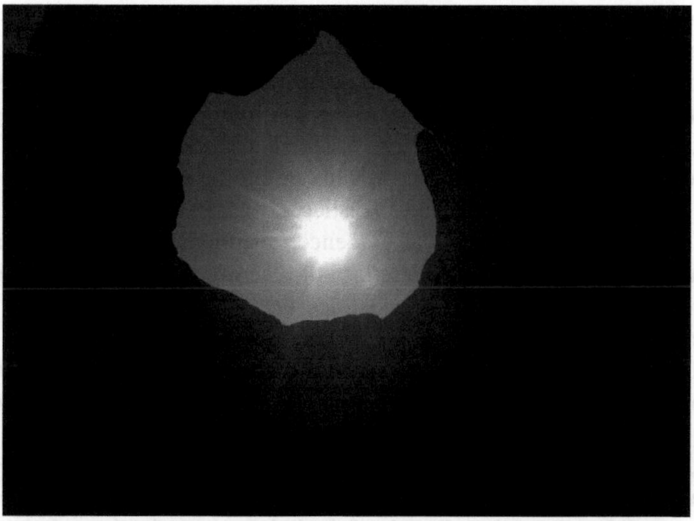

Fig. 3 Dawn of the winter solstice at the Campanaru of Monte Arcivocalotto

The imposing megalith on the side of Monte Arcivocalotto, just below the top of the hill, is visible even from a large distance. It is called in local dialect "u Campanaro" ("The Bell tower"). It is a large slab of arenite, about 4 m long and 3 m high. Its cross section is roughly triangular, with a thickness of about 1.5 m at the base; the western side is almost vertical, while the eastern one is inclined by about 75° with respect to the horizontal plane. At its center a circular, clearly artificial, hole is dug, with a diameter of about 2 m. In the internal thickness of the hole, a concave seat has been formed, of about 20 cm in height. A narrow step, about 50 cm wide, follows the northwestern side of the megalith, almost on the edge of the steep cliff of about 20 m in height over the small plateau, where the Bronze Age settlement was sited. On this step, just below the hole of the megalith, a petroglyph in concentric squares is carved.

Measurements made by a precision optical bearing compass[6] showed that the axis of the central hole has an azimuth of 133° ± 1° and a height from the horizontal plane of 15° ± 1°: this axis is clearly identifiable, even at a considerable distance, by an inverted V shaped notch in the upper edge of the hole. Because of this orientation, the sun rises over the geographic horizon at the center of the hole, exactly at the winter solstice (Fig. 3).

The megalith shows therefore a manifest astronomical orientation and there is clear archaeological evidence of intentionality: the hole and the inverted V notch, indicating the axis, are in fact clearly intentional. In addition, the megalith shows clear signs of works performed to point the axis of the hole in the selected direction.

[6]Measurements were corrected for local magnetic declination by detecting coordinates with GPS and comparison with the IGM cartography and geo-referenced satellite images.

Furthermore, the concentric squares petroglyph, engraved on the step at the foot of the megalith on its NW side, is aligned exactly as the axis of the hole: the petroglyph too is thus oriented towards the dawn of the winter solstice. On this day, and only on this day, the first rays of the rising sun illuminate the square engraved on the rock, because of two small notches carved into the bottom edge of the hole. Therefore, the petroglyph allows those who are on the step to recognize with great accuracy the exact day of the winter solstice.

With regard to the ethnological evidence, testimonies of the modern folklore show that the megalith was considered, until recently, a sacred and magical place, as is evidenced by the legend that justifies the megalith name of "Campanaru", identifying him as the bell tower of a destroyed church, which plays alone on special days. This legend then assigns the megalith mystical and calendrical role.

It remains to assess the statistical significance of this alignment. The probability of a single solar alignment in azimuth is equal to 1/22, corresponding, in Gaussian statistics, to 2.08 σ (Schaefer 2006). However, in the case of the megalith of Monte Arcivocalotto, we must consider that the phenomenon occurs only because the rising sun, as it passes to the azimuth of the hole center, also has a height over the astronomical horizon equal to the one of the geographical horizon, blocked for 15 degrees by the Rocca Busambra mountain: it can be calculated that the probability of this independent event is equal to 1/45 (Curti et al. 2009), corresponding, in Gaussian statistics, to about 2.5 σ. The alignment in azimuth and height with a significant solar position then has a conditional probability of random coincidence of 3.25 σ. In addition, the petroglyph axis azimuth alignment with the dawn of the winter solstice has an additional probability equal to 2.8 σ. The conditional probability of the simultaneous presence of these events is therefore equal to 3.56 σ and is thus higher than the internationally accepted threshold of 3 σ. We can therefore reasonably claim that the "Campanaru" satisfy all the conditions of the "Schaefer (2006) test" to be considered a megalith intentionally oriented in an astronomical direction.

3 The Relationship Between the "Campanaru" and Pizzo Pietralunga

It should also be considered that even the Pizzo Pietralunga site has other alignments with the dawn of the winter solstice.

Indeed, on this day, at the same time in which the rising Sun appears in the hole of the megalith, it is seen shining right behind Pizzo Pietralunga, when viewed from the right position. Of course, this is a purely natural event and therefore of no statistical value. However, at the same time, the rising Sun appears, from the same location, at the center of the image of Campanaru reflected in the river at the foot of Pizzo Pietralunga. Since the position on the hill slope of the megalith is not a natural fact, even for this event, clearly independent from those relating to the

Fig. 4 The rising sun at dawn of winter solstice is seen from Pizzo Pietralunga reflecting on the Belice river

megalith geometry, must be assigned a 3.25 σ probability of being random, involving an alignment in azimuth and height with the rising sun over the geographic horizon at the dawn winter solstice (Fig. 4).

In addition, a group of four artificial hole are presents in the Pizzo Pietralunga nearly vertical slope facing to South East. They were clearly made to house posts, which supported an unknown structure, probably a panel of some perishable material. These holes are aligned in such a way that at the dawn of the winter solstice, the sun shines directly on their bottoms. This means that at that time, and only at that moment, the shadow of the structure supported by posts was projected onto the rock surface exactly on the rectangle identified by the poles, giving a visible signal to the date of the solstice. Also to this event, independent of the previous ones, must be attributed a probability of randomness of 3.25 σ, as it also implies the coincidence azimuth and altitude of the holes axis with those of the rising Sun at the dawn of winter solstice.

On the other hand, it can be shown that the Campanaru and Pizzo Pietralunga are connected. In fact, an engraved stone was found near the megalith of Monte Arcivocalotto in 2012 (Fig. 5). On it, many symbols are engraved, the study of which is in progress. However, it is clear that the phallic symbol represented on the left repeats exactly the Pizzo Pietralunga profile. In addition, near the megalith, a worked monolith was found. It was felled, but intact and still in the original

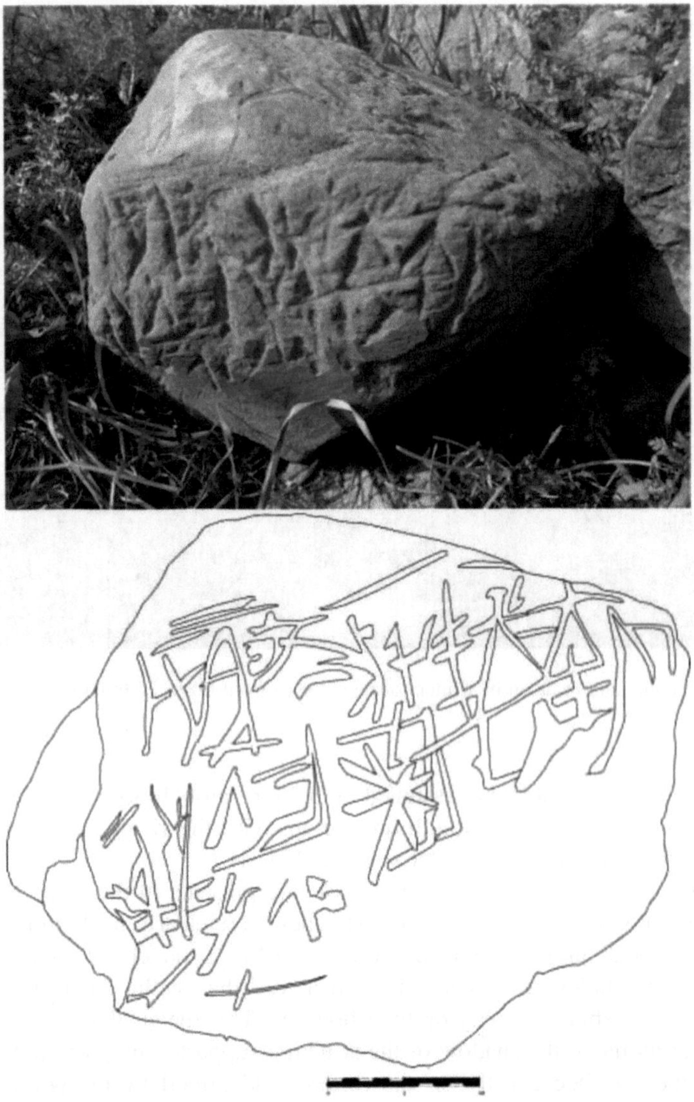

Fig. 5 The engraved rock found near the Monte Arcivocalotto megalith

location, exactly on the axis of Campanaru: working has made that this monolith has taken exactly the profile of Pizzo Pietralunga (Fig. 6).

In conclusion, these findings, along with clear archaeological evidence of contemporary attendance of the Monte Arcivocalotto and Pizzo Pietralunga sites by the same culture allow to claim that they are connected to each other. Therefore, the probability that the described simultaneous azimuth and height alignments to the dawn of the winter solstice are only occurring by chance is summarized in Table 1.

Fig. 6 The worked rock repeating the shape of Pizzo Pietralunga, straightened in its original position

Table 1 Probability of random coincidence of alignments with the dawn of the winter solstice in the Monte Arcivocalotto—Pizzo Pietralunga complex

Alignment to the dawn of winter solstice	Statistical significance respect to the null hypothesis (random alignment)—σ
Axis of the central hole and of the petrogliph in the *Campanaru* of Monte Arcivocalotto	3.56
Holes in the Pizzo Pietralunga slope	3.25
Sun rising in the reflected image of the *Campanaru* in the river at the base of Pizzo Pietralunga	3.25
Total compound probability	5.81

We recall that a Gaussian statistical significance of 5.81 σ corresponds to about one chance in a billion of chance coincidence. Given also the archaeological and ethnological evidence, we can thus reasonably claim to be in the presence of a megalithic complex intentionally oriented in an astronomical direction.

Other discoveries (Scuderi et al. 2013) reinforce this hypothesis. First, a few hundred meters from Pizzo Pietralunga a complex altar excavated in an isolated rock outcrop was discovered. The artifact is characterized by a stone staircase carved into the rock and fairly well preserved, which leads to the top of the boulder.

There, at various heights, basins collecting rainwater were dug: two of them are in communication by a hole, perhaps astronomically oriented. The study of this artifact is still ongoing, but its very presence confirms the importance of the Pizzo Pietralunga cult site. Furthermore, under the megalith of Monte Arcivocalotto, at the foot of the cliff that is located on its NW side, a tomb was discovered containing Early Bronze Age pottery fragments, including the ones of the "Bell Beaker Culture". Last, a group of four stones, probably artificially positioned, was discovered on top of Pizzo Pietralunga, by an ortho-photo taken from a kite. Their meaning is still under study.

4 The Cozzo Perciata Megalith

However, the most important discovery is certainly that of a second perforated rock (Scuderi et al. 2015). It is located about 6 km from Pizzo Pietralunga and 8 km from Monte Arcivocalotto, on top of a hill, significantly called Cozzo Perciata ("The hill of the drilled one"), a name that clearly refers to the existence of a perforated rock.

It has collapsed in recent times, it is said because a lightning, but many witnesses say that it was extremely similar to Monte Arcivocalotto "Campanaru", as evidenced by a photograph taken at the end of the '60s, or early '70s. This photographic image, by direct testimony of the author was attached to a file concerning the damage due to a drought episode. The photo depicts the Cozzo Perciata and the second perforated rock is very clearly and unequivocally visible on its top (Fig. 7). There is no reason not to believe original and certainly relative to Cozzo Perciata the photo, whose image is quite comparable to the present day profile of the Cozzo Perciata and shows, in the lower right angle of the photo, a stone water trough still existing at the southern foot of the Cozzo Perciata (Fig. 8).

Concerning the perforated rock, the lower part of the hole is still perfectly preserved, and this allows to easily measure the axis, with the same technique used for the megalith of Monte Arcivocalotto.

From the hole of the rock, the summit of Pizzo Pietralunga is viewed with an azimuth of 60.6° and a height of 1.7° (±1°): this direction corresponds to the dawn of the summer solstice over the geographic horizon in 2000 BC. Looking through the remains of the hole of the megalith of Cozzo Perciata at the dawn of the summer solstice, still today it is possible to see the rising Sun that touches the top of Pizzo Pietralunga. Then, from the bottom of the valley, everyone can see the Sun rising within what remains of the megalith (Fig. 9).

Important ethnographic testimonies of sacred and calendar value are reported even for this rock: it too was called in Sicilian *'u Campanaru* or even *'a petra unni nasci u suli* ("The stone where the Sun rises"). The same witnesses who have indicated these names state that, until a few decades ago, the rise of the Sun in the hole of this rock was the signal of the beginning of the period for harvest.

Fig. 7 The old picture showing the Cozzo Perciata megalith: **a** original picture; **b** detail of the megalith

Fig. 8 A present day picture of Cozzo Perciata. The *upper arrow* shows the perforated rock; the *lower one* the water trough

As for archaeological evidence in the Cozzo Perciata area, already in 1997 some ceramic fragments were published (Scuderi et al. 1997, pp. 333–340), of certain attribution to Eneolithic and Early Bronze Age, in particular to the San Cono-Piano Notaro and Rodì-Tindari-Vallelunga facies (Fig. 10).

More recent surveys, carried out both on the ridge and on the plateau to the Southeast, have led to the discovery of flint and obsidian blades (Fig. 11) and pottery fragments dating from the Late Neolithic to the Bronze Age and occasional fragments attributable to the Iron Age. The dispersion area of the fragments is about 5,000 m^2, although a large part of the archaeological area suffered utilization of agricultural type in other historical periods.

It is worth to remind a fragment of the Bell Beaker Culture (Fig. 12), and an innumerable amount of fragments belonging to the pre-Castellucciana and Castellucciana cultures. The same typology of ceramic fragments were found in a tomb discovered under the pierced rock: the Monte Arcivocalotto, Pizzo Pietralunga and Cozzo Perciata sites are, to date, the only area in which there is the presence of ceramics attributable to the Bell Beaker Culture that, in other European regions, is indisputably linked to Megalithism. The analogy between the megaliths of Monte Arcivocalotto and Cozzo Perciata is therefore very strong.

Recently, less than 1 km from Cozzo Perciata, a roughly hemispherical shaped artificial cavity (about 5 m diameter × 3.50 height) with a circular summit hole,

Fig. 9 Sunrise at the summer solstice at Cozzo Perciata: **a** the firsts rays of the rising Sun touch the top of Pizzo Pietarlunga (shown by the *white arrow*); **b** the rising Sun shine at the centre of the remains of the hole; **c** the Sun rises at the position of the perforated rock

Fig. 10 Eneolithic and Early Bronze Age ceramic fragments (San Cono-Piano Notaro and Rodì-Tindari-Vallelunga facies) found in Cozzo Perciata

Fig. 11 Flint and obsidian blades found in Cozzo Perciata

Fig. 12 Fragment of Bell Beaker Culture ceramics found in Cozzo Perciata

Fig. 13 **a** The entrance of the tholos tomb; **b** the access corridor of the tomb; **c** the pinnacle and the cup-mark above the tomb

now mostly artificially walled, has been discovered. The cavity is reached through an open-air corridor also cut into the rock and a room-door roughly rectangular. Outside, above the cavity, a pinnacle and a cup-mark carved into the rock are present (Fig. 13).

The typology, and in particular the presence of the dromos carved into the rock, leads immediately and with little doubt to the Thapsos tombs with dromos and, in closer geographical context, to the tholoid tombs of Sant'Angelo Muxaro and of the area of Platani (Tommasello 1997).

Ultimately, the fact that, in this area we found two coeval and similar monuments (rocks artificially perforated) with solstice alignments in azimuth and height different and complementary (winter and summer solstices) shows that here, between the Eneolithic and the early Bronze Age, a culture has developed that had a solar calendar and developed a simple, but very effective, technology to materialize it.

5 The “King’s Pulpit”

Another monument gives us further guidance on what might have been this culture. The rocky artifact, known locally as “Pulpito del Re” (“The King’s Pulpit”[7]), is located in the province of Palermo (geographical coordinates 37° 52′ 53″ N, 13° 24′ 00″ E, 864 m height above sea level), inside the natural park of the “Bosco della Ficuzza”, not far from the megalithic complex centered on Pizzo Pietralunga (Scuderi et al. 2014). The geographical area, in which this artifact is situated, has been little studied so far from the archaeological point of view, although it is located in a territory whose attendance is documented from Eneolithic to all the Early Bronze Age.

In particular, not far from the pulpit, our group has discovered a settlement dated, by the large number of ceramic fragments on this site, to the Castellucciano (a culture of the initial phase of the Early Bronze Age, especially widespread in the South-East Sicily and temporally placed between 2200 and 1400 BC). Near this settlement, on the path leading to the pulpit, we found a flat rock, more or less oriented towards North-South, where the image of a bull is engraved. A basin carved on it and many cup-marks on the bull body suggest a sacred role of this rock (Fig. 14).

The King’s Pulpit is made from a natural outcrop of siliceous rock, with an approximately circular base (Fig. 15). A staircase carved on this rock leads to a semicircular clearly artificial platform, which represents a kind of “podium”, with a flat base, surrounded by seats with backrest, also carved in the same rock. The staircase consists of 12 steps, is 6.4 m long and has a width of 0.75 m. It is inclined

[7]The popular name, reported only in the Twentieth Century maps, is explained locally by the legend that it was built as a hunting station for King Ferdinand IV of Bourbon (1759–1825), whose luxurious summer residence was actually built a few kilometers away, at the beginning of the Nineteenth Century. However, this explanation has no foundation, because the artifact is shown on a map of the early Eighteenth Century, thus more than half a century before King Ferdinand IV, as “a rock engraved with a staircase”. Moreover, the upper part of the artifact is clearly visible from all sides and is therefore unsuitable as a hiding place for hunting. However, ethnographic studies have not been done on this name and its origin, so we do not know how much it is ancient: we can not exclude that it was named so since a long time and was later associated with Ferdinand IV. Indeed, at about 200 m from the pulpit, there is another rock modified by man, similarly shaped in some respects, but definitely of much more recent work, as suggested by the lack of erosion processes. Even in this case, a staircase has been excavated on the rock, leading to an upper base, where the Bourbon crest is graved. Furthermore, this other rock is placed in an elevated position, but sheltered from view, which seems much more suitable, than that of the “King’s pulpit”, as “hunting station”. It is thus reasonable to assume, that this is the true pulpit of King Ferdinand IV, and that, with the passage of time, a confusion between the two artifacts occurred.

Fig. 14 The bull engraved in the rock

Fig. 15 The Pulpit of the King top "podium"

by 22° and presents a kind of groove, which may be due to repeated dragging of a heavy load. In addition, all the rock shows a series of basins, niches, cup-marks, reticular carvings and perforations of obvious artificial origin. Many ceramic fragments of the Early Bronze Age were recovered near the pulpit: this fact and the same structure of this artifact allow to identify it as a Castellucciano altar, in many respects similar to those of Rocca Pizzicata and Pizzo Pietralunga.

A rocky protuberance has been left at the center of the podium and a clearly artificial hole was carved in front of it, in the semicircle of the rear part of the seats

Fig. 16 The Pulpit and the second rocky spur

(Fig. 16). This approximately conical bore, of irregular shape, pass through all the rock up to its exterior, where the longest side is of about 30 cm and the maximum height is of about 10 cm. Over the hole, a V-shaped incision is carved into the rock: this incision, the hole and the tip of the protuberance on the platform lie in the same vertical plane.

In a second rock outcrop, 10.5 m from the pulpit, another notch in the shape of V is artificially carved. This incision, the tip of the protuberance and the hole in the back of the pulpit seats are aligned on the same axis.

The Pulpit orientation measurements were carried out by us on 20th December 2013 and 9th May 2014, using a precision bearing compass (Optical compass DQL-6 model). The magnetic declination was determined by measuring the position of the Sun with the same instrument (with the objective shielded by a Hα filter) compared with astronomical ephemeris. It was found equal to 2.5° ± 0.2° E, a value in good agreement with the NOAA model, which gives, for the geographical coordinates of the site (measured with a GPS Magellan eXplorist GC model, with an instrumental uncertainty translated into linear units of ±20 m) and the measurement date (May 9th 2014) a value of 2.53° E.

These measurements have shown that the orientation of the staircase (geographical azimuth = 180.5° ± 0.5°) perfectly corresponds to the local meridian.

The sight line connecting the hole, the tip of the central protuberance on the podium and the V-shaped incision on the second rocky spur placed in front of the Pulpit is more difficult to measure, given that the irregular shape of the hole makes it difficult to identify a precise point of observation and the protuberance is too low to allow the positioning of the optical compass. For this reason, this line of sight was measured in the opposite direction, i.e. from the incision in the second rock to the tip of the protuberance and the hole. It has been found that this axis has a geographic azimuth of 60° ± 0.5° and a height, with respect to the horizontal plane of −4.4° ± 0.5°. The astronomical declination of the opposite viewing direction therefore corresponds to 22.8° ± 9.9°, being the uncertainty due to the error propagation of the azimuth and elevation uncertainties in the declination formula. The central value of the declination, however, coincides almost exactly with that of

Fig. 17 The hierophany of the summer solstice dawn at the King's Pulpit

Fig. 18 The pinnacle of the third rock is illuminated by the Sun rays passing through the hole in the Pulpit

the sun at the summer solstice: it was thus decided to observe the possible effects of light on the structure at the dawn of this date.

One of us (AS) was therefore present at the pulpit at sunrise on June 21st, 2014. It was observed that the Sun actually rises inside the hole, it shines in its center after a few minutes and then appears in the V-shaped incision in the back seat over the hole (Fig. 17). At this time, the Sun rays illuminate a rocky artifact on the back of the pulpit, where a copy of the protuberance on the podium is carved (Fig. 18). The probability that such a hierophany happens by chance is very low. In fact, considering that each alignment in azimuth and elevation with one of the 8 significant solar positions has statistical significance of 3.3 σ compared to the null hypothesis and that there are in this artifact 4 of these alignments[8] (Fig. 19) plus a coincidence azimuth (scale of the Pulpit—meridian) which involves other 2.08 σ of significance, the resulting composed statistical significance is 6.9 σ.

The Castellucciano altar known as the King's Pulpit shows thus clearly that the astronomical alignments were deliberately sought in this artifact. The amazing hierophany happening at sunrise on the summer solstice seems to be designed for a complex ceremony to be performed on that day.

[8]Hole—viewfinder in the second the rock, hole—pinnacle on the Pulpit podium, pinnacle on the Pulpit podium—viewfinder in the second rock, viewfinder on the seats of the Pulpit—pinnacle in the second artifact.

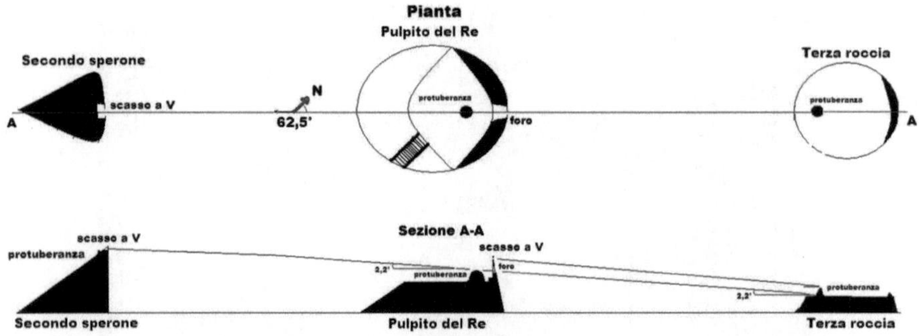

Fig. 19 Schematic plant and section of the King's Pulpit complex

6 Conclusions

In conclusion, sufficient objective data support the hypothesis that South of Monte Jato we are faced with significant archaeological discoveries with obvious astronomical implications. The two perforated rocks, located at a distance of less than 8 km from each other, respectively indicate the dawn of the winter solstice and the summer solstice, with all the cultural, agricultural and worship implications that can be consequently hypothesized. The intentionality and the connection of the two monuments can hardly be questioned. Hypothetically one can imagine that, to achieve this technological result, careful and long, perhaps even years, research has been accomplished of two rocks in the same area having the characteristics and the desired orientation, before proceeding with appropriate skills to their processing to obtain the desired effect.

It would appear superfluous to recall here, following Eliade (1955, Chap. VI), the importance of the role that the artificial or natural "pierced stone" play in anthropology and ethnography. The same popular memory of the Palermo "Emperor stone", whose shadow, in the time of Frederick II, would indicate the end of the working day for agricultural laborers, constitutes testimony, though faded, of an ancient use of alignments of natural or artificial stones and rocks and the shadow they cast, to mark time.

Reasoning on the age in which the two large holes were opened in the respective rocks, with particular purposes related to the solar position at the time of the solstice sunrises, is obviously not easy. However, we believe that one can assume with good reason the existence in prehistoric times of a large area, an "archaic landscape" with a great sacred value in the territory South of Monte Jato.

The extraordinary hierophany of the summer solstice, made following similar principles (perforations in natural rocks) but with far greater sophistication, at the Castellucciano altar named "Pulpit of the King" makes us lean towards the hypothesis that the Castellucciana culture is also the author of "stone calendars" of Monte Arcivocalotto and Cozzo Perciata. The astronomical orientation of other artifacts of the same culture therefore deserve further studies.

References

Caracausi, C.: Dizionario onomastico della Sicilia, 2 vol. Epos, Palermo (1994)

Curti, E., Mucciarelli, M., Polcaro, V.F., Prascina C., Witte, N.: The "Petre de la Mola" megalithic complex on the Monte Croccia (Basilicata). In: Shaltout, M., Rappenglueck, M. (eds.) Proceedings of the SEAC 17th Annual Meeting from Alexandria to al-Iskandariya, Astronomy and Culture in the Ancient Mediterranean and Beyond. B.A.R., London (2009)

Eliade, M.: Trattato di storia delle religioni. Edizioni scientifiche Einaudi, Torino (1955)

Nania, G.: Toponomastica e topografia storica delle valli del Belice e dello Jato. Bardaro Editore, Palermo (1995)

Polcaro, V.F., Scuderi, A., Mercadante, F., Lo Cascio, P.: U Campanaru: un monumento per la misura del tempo? Archeologia Viva **156**, 50–57 (2012)

Scharfer, B.E.: Case study of three of the most famous claimed archaeoastronomical alignments in North America. In: Bostwick, T.W., Bates, B. (eds.) Proceedings of Oxford VII International Conferences on Archaeoastronomy Viewing the Sky Through Past and Present Cultures, Pueblo Grande Museum Anthropological Papers no. 15, Phoenix (AZ), pp. 27–56 (2006)

Scuderi, A., Tusa, S., Vintaloro, A.: La preistoria e protostoria del territorio di Corleone nel quadro della Sicilia occidentale, Corleone (1996)

Scuderi, A., Tusa, S., Vintaloro, A.: La preistoria e protostoria nel Corleonese e nello Jato, Corleone (1997)

Scuderi, A., Mercadante, F., Lo Cascio, P.: La valle dello Jato tra Archeologia e Storia, San Cipirello (2011)

Scuderi, A., Mercadante, F., Lo Cascio, P., Polcaro, V.F.: The astronomically orientend megalith of Monte Arcivocalotto. In: Šprajc, I., Pehani, P. (eds.) Ancient Cosmologies and Modern Prophets. Proceedings of the 20th Conference of the European Society for Astronomy in Culture, Ljubjana, pp. 213–221 (2013)

Scuderi, A., Vitale, F., Tusa, S., Polcaro, V.F., Scuderi, R., Maurici, F., Leone, I.: Il Pulpito del Re. In: Giritto, V., Rosada, G. (eds.) il Cielo in terra ovvero della giusta distanza, Proceedings of the 14th Conference of the Italian Society for Archaeoastronomy, pp. 145–152. Padova University Press (2014)

Scuderi, A., Polcaro, V.F., Mercadante, F., Lo Cascio, P., Maurici, F.: The astronomically oriented megaliths of the Monte Jato area (Sicily): the "Campanaru", the "Perciata" and the Eneolithic/Early Bronze Age worship site of Pizzo Pietralunga. Mediterr. Archaeol. Archaeometr. **14**, 156–165 (2015)

Tomasello, F.: Le tombe a tholos nella Sicilia centro meridionale, Cronache, XXXIV–XXXV, 1995–96, Catania (1997)

The First Archaeoastronomical Study of the Maltese Temple of Borġ In-Nadur

Andrea Orlando and Davide Tanasi

Abstract The Maltese island has megalithic temples of extraordinary interest for archaeoastronomy. In literature we find different works that involve most of its archaeological sites. The temple of Borġ in-Nadur, set on the top of a hill by the Marsaxlokk Bay in southern Malta, is less well known than the rest of the others, even though it started off as a major attraction for grand tourists and travellers in the Early Modern and Colonial periods. It was explored in the second half of the 1920s by a team of British archaeologists, led by Margaret Murray, who gradually uncovered the ruins of typical Maltese megalithic temple dated to the 3rd millennium BC. In 2011 the efforts of an international team of scholars brought to the publication of a general reassessment of the evidences about the temple of Borġ in-Nadur and the artifacts collected during its exploration, emphasizing its importance for the Maltese Temple Period. The new picture that emerged has reactivate the research around the Borġ in-Nadur temple attracting for the first time the interest of scholars in archaeoastronomy. The archaeoastronomical study of the Borġ in-Nadur's archaeological site is the first of its kind, as the archaeological remains were put into evidence in 2010, and pays particular attention to the temple. At the moment it is difficult to determine the exact plan of the entire temple, and therefore its axes, but the apsidal building and the main entrance are quite intact. Although this research is at its early stage of development, with regards to archaeoastronomical issues and features related with the temple of Borġ in-Nadur some preliminary considerations can be put forth.

A. Orlando (✉)
Catania Astrophysical Observatory (INAF), Catania, Italy
e-mail: aorlando@oact.inaf.it

A. Orlando
National South Laboratories (INFN), Catania, Italy

A. Orlando
Institute of Sicilian Archaeoastronomy (IAS), Novara di Sicilia, Italy

D. Tanasi
Center for Virtualization and Applied Spatial Technologies (CVAST) & Department of History, University of South Florida, Tampa, FL, USA

A. Orlando (ed.), *The Light, The Stones and The Sacred*, Astrophysics and Space Science Proceedings 48,
DOI 10.1007/978-3-319-54487-8_3

1 Introduction

About 20 megalithic sites are known in Malta and Gozo and together they probably represent the most relevant tourist attraction of the archipelago and the backbone of the Maltese archaeological scientific literature. However, the temple of Borġ in-Nadur, set on the top of a hill by the Marsaxlokk Bay in southern Malta, is less well known than the rest, even though it started off as a major attraction for grand tourists and travellers in the Early Modern and Colonial periods. It was explored in the second half of the 1920s by a team of British archaeologists, led by Margaret Murray, who gradually uncovered the ruins of typical Maltese megalithic temple dated to the 3rd millennium BC.

The excavations uncovered a monumental sacred complex, characterized by a singular plan including a megalithic enclosure with different cult places. A large number of finds were unearthed, demonstrating the wealth of the community using the site. At that time, the conditions of the temple building were rather good. The preliminary reports of the explorations, published promptly in 1923, 1925, and 1929, were accompanied by a thorough drawn and photographic documentation including an accurate measurements of nearly all megaliths (Murray 1923, 1925, 1929).

2 The Temple of Borġ in-Nadur

The site of Borġ in-Nadur is set on a hill on the St. George Bay, and is comprised of two different settlements: the Bronze Age fortified village on the top of the hill (Tanasi and Vella 2015) and the megalithic temple on the eastern slope that was in part reused and modified throughout the Tarxien and Borġ in-Nadur periods (Tanasi and Vella 2011) (Fig. 1).

Murray's investigations revealed the following remains: an Apsidal Building, an Open Area or Main Enclosure, the Field Stones and a Double Chapel (Figs. 2, 3 and 4). The Apsidal Building is a four-apsed temple with a shallow niche at the end. The Open Area or Main Enclosure lies outside the temple and a good part of the megalithic wall was uncovered by excavation. The Apsidal Building can be safely identified as a 'temple' on the basis of the similarities in ground plan with other sites. The megalithic set-up of the Open Area, then, could have defined the temple forecourt already in the Temple period. The arrangement is not dissimilar, in fact, to what artists recorded beneath the temple complex at Ġgantija, Gozo, in the nineteenth century. The façade of the Main Enclosure was marked by the sequence of four elements, a monumental main gate, a great menhir and a dolmen (Fig. 5).

Along the northern wall of the Main Enclosure, the fourth orthostat shows a large e perfectly circular artificial hole passing through it and a triangular niche with three half spherical sockets carved on the outer surface (Fig. 6).

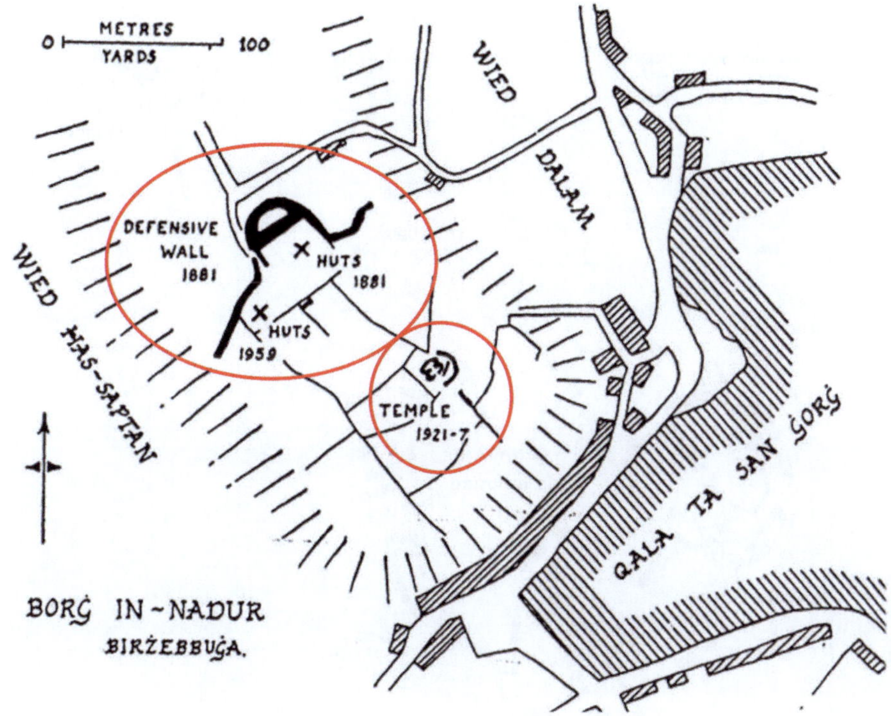

Fig. 1 Plan of Borġ in-Nadur site with indication of the Bronze Age settlement and of the megalithic temple (after Trump 1961)

Another smaller enclosure very poorly preserved, named Field Stones, was probably located to the North of the main complex, and a second apsidal building closed by a smaller precinct, called the Double Chapel, was explored to the South-East. Groups of objects including pottery sets, loom weights, and also an anthropomorphic betyl idol (Fig. 7), found in different areas of the Double Chapel and related to Bronze Age period may suggest cultic activities still active during the Bronze Age at least in this part of the complex.

In the past 80 years, for different reasons this site was forgotten and generally neglected with the result that the current conditions of the entire archaeological area are unfortunately rather poor.

Due to the phases of occupation subsequent to Temple period, many megalithic elements were removed from their original positions and completely de-functionalized and decontextualized. That especially affected betyls, menhirs and the so called notched stones (Figs. 8 and 9).

This has jeopardized the interpretation of the general organization of the complex and as a consequence, Borġ in-Nadur has not been included in any tourist itinerary and its role in Maltese prehistoric has been always considered marginal.

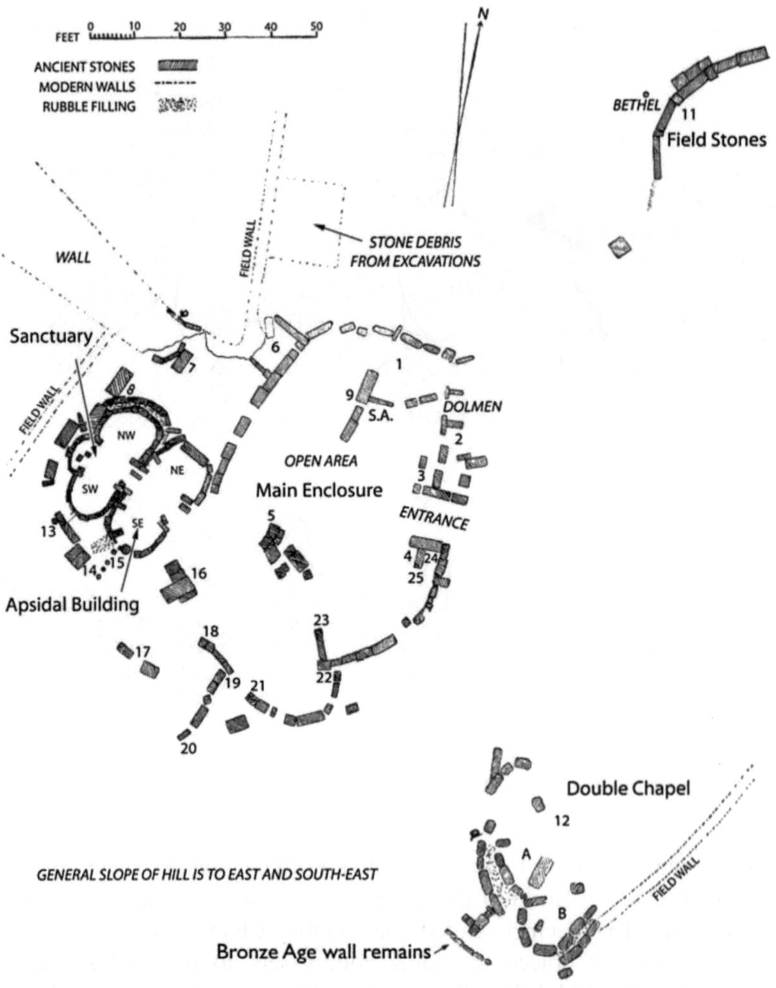

Fig. 2 General plan of Borġ in-Nadur temple (Tanasi and Vella 2011)

Against this scenario, in 2011 the efforts of an international team of scholars brought to the publication of a general reassessment of the evidences about the temple of Borġ in-Nadur and the artefacts collected during its exploration (Tanasi and Vella 2011), emphasizing its importance for the Maltese Temple Period. The new picture that emerged has reactivate the research around the Borġ in-Nadur temple attracting for the first time the interest of scholars in archaeoastronomy, a discipline which has recently proved to have a fertile field of application in prehistoric Malta (22th SEAC Conference, Malta 21–27 September 2014).

In particular, the production of a virtual interactive reconstructive model in 3D computer animation of the temple has resulted instrumental for the reactivation of the research about this monument (Tanasi and Stanco 2013). The temple has been

Fig. 3 The temple at the time of Murray's excavations (Murray 1929)

Fig. 4 The Apsidal Building and the Main Enclosure from West (photo authors)

rebuilt using the measurements provided in Murray's reports, while all the other structures were reconstructed using dimensions recorded on site or through comparisons with other temple sites (Figs. 10, 11 and 12).

The reconstruction work has been completely carried out using the Blender 3D suite,[1] an open source cross-platform software for modeling, rendering, animation, postproduction, creation, and playback of interactive 3D contents. The 3D model was not intended to reconstruct in elevation the missing parts of the temple but was aimed at rediscovering digitally what was found by the archaeologists nearly 80 years ago. In order to add realism to the digital replica, a study of light sources

[1]Visit the website: http://www.blender.org.

Fig. 5 Detail of the outer wall of the Main Enclosure, from East: (*1*) main gate; (*2*) the great menhir; (*3*) the broken dolmen (photo authors)

Fig. 6 The fourth orthostat of the outer wall of the Main Enclosure (photo authors)

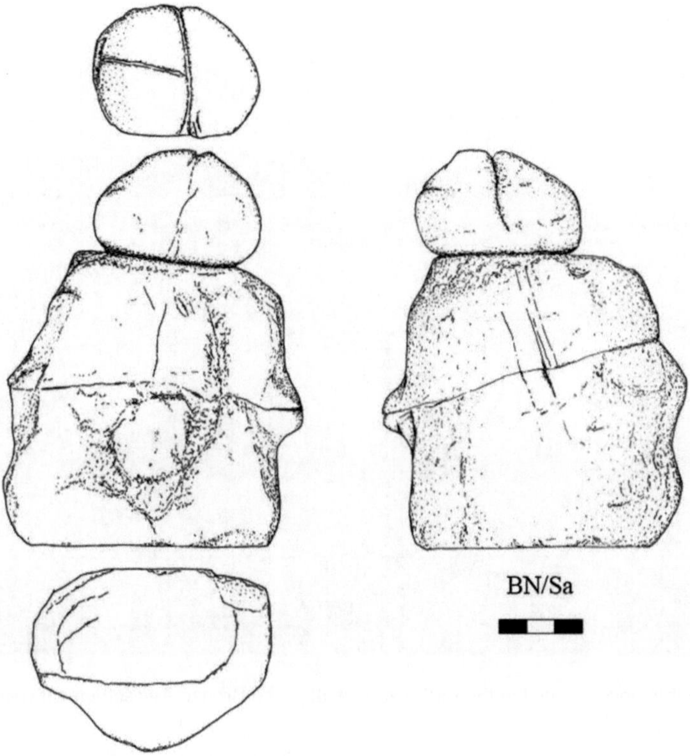

Fig. 7 Betyl idol from the Double Chapel (Veca 2011)

was carried out, simulating a complete cycle of the sun on mid-summer day through the use of the Radiance raytracer[2] (Fig. 13).

3 Archaeoastronomy in Malta

The Maltese island has megalithic temples of extraordinary interest for archaeoastronomy. In literature there are several works about such sites. Among them, it must certainly be remembered the contributes of Agius and Ventura (1981) and Foderà Serio et al. (1992). Agius and Ventura's was the first work considering the possible astronomical alignments of the Maltese temples. The study applied a

[2]Visit the http://radsite.lbl.gov/radiance.

Fig. 8 Menhirs incorporated in the fortification wall of the Bronze Age settlement (photo authors)

maximal approach: for axis of the temple it was considered just what was clearly a feature of the construction, even if it was not a true axis of symmetry.

Out of 26 azimuths, there was a clear concentration of 20 of them between 128° and 230°, between SE and SO, while the remaining 6 showed no patterns (Fig. 14).

In their analysis of the data, the authors of the first work considered the rising and setting positions of the Sun at solstices and equinoxes and of the Moon at standstills. Moreover they studied the alignment of a number of temples towards the rising or setting positions of some bright stars: Sirius, α Crucis, α Centauri and β Centauri, α Lyrae and α Ursa Majoris.

In the work of Foderà Serio et al. (1992), it was applied instead a minimal approach: for axis of the temple it was considered just what was an unequivocal axis of symmetry. In this case just 15 azimuths were measured (Fig. 15), and the authors took into account mostly the possible alignments of the temples at sunrise or setting positions of some bright stars: Sirius, Canopus, α Crucis, β Crucis and γ Crucis, α Centauri and β Centauri, and The Pleiades.

Fig. 9 Fragment of notched stone re-employed for the construction of a later wall (photo authors)

Fig. 10 Virtual interactive model of the Apsidal Building (Stanco and Tanasi 2013)

Fig. 11 Virtual model of the Main Enclosure (Stanco and Tanasi 2013)

Fig. 12 Virtual model of the fourth orthostat of the outer wall of the Main Enclosure (Stanco and Tanasi 2013)

Fig. 13 Virtual study of the effect a solar cycle on the shadows produced by the megaliths of the Borġ in-Nadur temple (Stanco and Tanasi 2013)

4 Archaeoastronomy at Borġ in-Nadur: Preliminary Analysis

Due to the scarcity of previous data, the archaeoastronomical study of the temple of Borġ in-Nadur is the first of its kind and obviously the study is in its infancy. Without a proper survey and technical analysis on-site, it is difficult to determine the exact plan of the entire temple and in particular its axes, although but the Apsidal Building and the main entrance are quite intact. From an initial examination with satellite data, it was found that the azimuth of the apsidal building is 125°, then framed in the angular range determined by Agius and Ventura, while the azimuth of the main entrance is 107° (Fig. 16). The temple of Borġ in-Nadur has several characteristics that make it unique, as the many standing stones and the peculiar location of the dolmen in the wall of the Main Enclosure.

The first step in the research agenda will be a fieldwork already scheduled for Spring 2018, aimed to acquire all the missing technical data and to start comparing them with those already available for the other megalithic temples. Using the technological support of the Center for Virtualization and Applied Spatial Technologies (CVAST) of University of South Florida, a terrestrial and aerial LIDAR scanning of the Borġ in-Nadur temple will be carried out.

3D scans will be georeferenced and complemented as needed by digital photogrammetry, and the assembly of all digital georeferenced data will be built into a GIS environment. Laserscanning data will be collected at a nominal minimal resolution of 5 cm across the entire study area and areas of special interest will be collected at a nominal minimal resolution of 1 cm. Digital photogrammetric

Temple		Axis [a]	Azimuth	Declination	
Bugibba		1 - 2	201° [b]	-49.5°	S [e]
Hagar Qim	I	3 - 4	128.7°	-30.5°	R
Hagar Qim	I	4 - 3	308.7°	30.5°	S
Hagar Qim	I	5 - 6	297.8°	23.3° [d]	S
Hagar Qim	I	7 - 8	0.8°	54.4° [d]	R
Hagar Qim	I	9 - 10	213° [c]	-42.8°	S
Hagar Qim	I	11 - 12	255 ° [c]	-12.1°	S
Hagar Qim	II	13 - 14	186.0°	-53.7°	S
Ta' Hagrat	I	15 - 16	130.6°	-29.4° [d]	R
Ta' Hagrat	II	17 - 18	172.7°	-48.7° [d]	R
Ggantija	I	19 - 20	128° [b]	-29.9°	R
Ggantija	II	21 - 22	133° [b]	-33.5°	R
Kordin	I	23 - 24	149.6°	-44.3°	R
Kordin	II	25 - 26	199° [c]	-50.0°	S
Mnajdra	I	27 - 28	92.7°	0.0° [d]	R
Mnajdra	II	29 - 30	138.1°	-38.1° [d]	R
Mnajdra	III	31 - 32	207.2°	-46.1°	S
Tal-Qadi		33 - 34	76°	11.3°	R
Skorba	I	35 - 36	134.9°	-33.3°	R
Skorba	II	37 - 38	168.5°	-50.1° [d]	R
Tarxien	I	39 - 40	200.2°	-49.5°	S
Tarxien	II	41 - 42	230.1°	-31.3°	S
Tarxien	II	43 - 44	142.0°	-39.7°	R
Tarxien	III	45 - 46	198.7°	-50.1°	S
Tarxien	IV	47 - 48	170° [c]	-52.9°	R
Xrobb il-Ghagin		49 - 50	140° [b]	-38.4°	R

(a) Refer to the plans for identification of the axes;
(b) Estimated from published plans;
(c) Estimated from plans whose reliability was checked against measurement;
(d) The altitude of the horizon and refraction taken into consideration;
(e) R = rising and S = setting positions

Fig. 14 The table shows azimuth and declinations of the temples' axis (from Agius and Ventura 1981)

products will be produced at comparable resolutions. Georeferencing data as needed will be collected using dGPS to ensure the locational accuracy of the data. dGPS data will have a minimal accuracy of 0.2 m. All data will be transformed into UTM projections. Deliverables will be registered and georeferenced point clouds for each study site together with appropriate metadata and a report detailing the project plan and workflow. Point clouds will be delivered in the .las or .laz format.

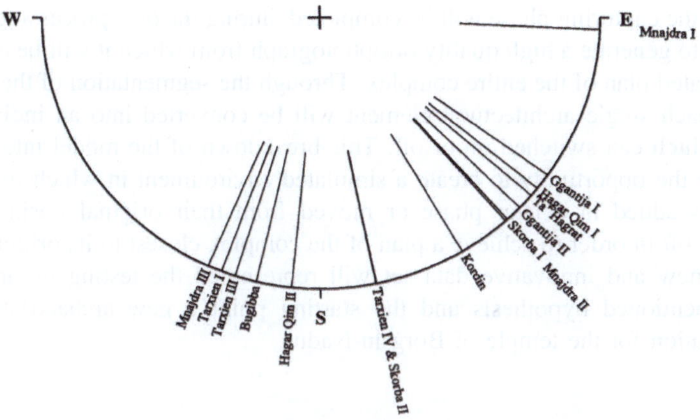

Fig. 15 The orientations' diagram with the azimuth of the temples' axis (Foderà Serio et al. 1992)

Fig. 16 Borġ in-Nadur. Satellite view of Temple, with the Google Earth ruler showing the azimuth of the building (Image courtesy Google Earth, drawing by the authors)

The GIS will consist of publicly available satellite or aerial photographic base imagery, the georefenced point clouds, and annotations attached to vector features that will guide a user.

The Maltese site presents the same architectural features and setting of the Bronze Age site of Motilla del Azuer in Castilla-La Mancha (Spain), 3D captured by CVAST scientific team in 2016. The digital outcome, currently available for scholars and general public in the CVAST Sketchfab collection,[3] has an excellent level of quality that makes it a valuable interpretative tool.

---

[3]Visit the webpage https://sketchfab.com/models/820ff9597ee84d3996589c37acce38a7.

Once the capturing phase will be completed, during the data processing it will be possible to generate a high quality ortophotograph from which it will be extracted a new updated plan of the entire complex. Through the segmentation of the entire 3D model, each single architectural element will be converted into an individual 3D model which can switched on or off. This breakdown of the model into 3d layers will give the opportunity to create a simulated environment in which for example megaliths added in a later phase or moved from their original position can be switched off in order to achieve a plan of the complex closest to its original design.

This new and innovative data set will represented the testing ground for the above mentioned hypothesis and the starting point a new archaeoastronomical interpretation for the temple of Borġ in-Nadur.

5 Conclusions

The archaeoastronomical reappraisal of the temple of Borġ in-Nadur at first glance promises to provide interesting data which will prove once again the profound knowledge of the ancient Maltese people for the natural phenomena, the solar cycle and the stars. This study never carried out before already represents an important novelty that however just enlarge an existing data set of cases study.

A game changing novelty, instead, could come by the contextualization of a new archaeoastronomical evidence in the frame of the reuse of the Borġ in-Nadur in the Middle Bronze Age, where all around it was established a settlement surrounded by megalithic and gargantuan walls (Tanasi and Vella 2015). The outstanding defences of the temples, cannot be archaeologically explained as there are not traces of external threats of internal competing forces. In particular, the massive D-shaped bastion (see Fig. 1), does not have comparison in the megalithic architecture of the Bronze Age Mediterranean, from Syros in the Aegean to Los Millares in the Balearic archipelago. Would it possible to hypothesize that the ruins of the ancient temple empowered by its astronomical connections justified the plan and the development of the Bronze Age settlement? Would it be possible to infer that unique fortification without equals was meant to protect and keep safe those ruins? Finally would it be possible to explain the unusual feature of the D-shaped bastion calling in reasons connected to a lost astronomical knowledge?

These rhetorical questions, on one side, are obviously meant to provoke but on the other side they inform us about how complex the archaeoastronomical reappraisal of Borġ in-Nadur as site can become and long is still the way that takes us to the stars.

References

Agius, G., Ventura, F.: Investigation into the possible astronomical alignments of the copper age temples in Malta. Archaeoastronomy (The Bulletin of the Center for Archaeoastronomy, University of Maryland) **IV**(1), 10–21 (1981)

Foderà Serio, G., Hoskin, M., Ventura, F.: The orientations of the temples of Malta. J. Hist. Astron. **23**(2), 107 (1992)

Murray, M.A.: Excavations in Malta: Part I. Bernard Quaritch, London (1923)

Murray, M.A.: Excavations in Malta: Part II. Bernard Quaritch, London (1925)

Murray, M.A.: Excavations in Malta: Part III. Bernard Quaritch, London (1929)

Stanco, F., Tanasi, D.: Beyond virtual replicas. 3D modeling and Maltese prehistoric architecture. J. Electr. Comput. Eng. 1–7 (2013)

Tanasi, D., Vella, N.C. (eds.): Site, Artefacts, Landscape: Prehistoric Borġ in-Nadur, Malta. Polimetrica, Monza (2011)

Tanasi, D., Vella, N.C. (eds.): The late prehistory of Malta: essays on Borġ in-Nadur and other sites. Archaeopress, Oxford (2015)

Trump, D.H.: The later prehistory of Malta. Proc. Prehist. Soc. **27**, 253–262 (1961)

Veca, C.: The small finds. In: Tanasi, D., Vella, N.C. (eds.) Site, Artefacts, Landscape: Prehistoric Borġ in-Nadur, Malta, pp. 195–224. Polimetrica, Monza (2011)

The Criticisms of Claudius Ptolemy to Marinus of Tyre in the *Geographia* and the Geographical Data of the Meridian Line of St. Nicholas in Arenis, Catania (Sicily)

Nicoletta Lanciano and Eleonora Cicciarelli

Abstract The study presented was carried out as part of a larger project on ancient Geography and in particular on the geographical data of Claudius Ptolemy, directed to the criticisms that Ptolemy addresses to Marinus of Tyre. Ptolemy collects data provided by Marinus, such as the correspondence of 500 stadia for each Earth maximum circle degree, he uses the geographical methods used by Marinus, but criticises Marinus' cylindrical maps and making new ones, and he criticises the way of processing the observational data. Ptolemy's calculations are published, and the discussion arising from them is discussed, concerning the use of the cited stellar coordinates by considering the Precession and the coordinates of the geographic locations to which the data of the stellar astronomical observations (Tyre, Rhodes, Alexandria). In addition, a link is proposed between the geographical problems of Ptolemy and the reading of the great Meridian tangent of St. Nicholas in Arenis in the Benedictine complex in Catania, and of some of the geographical data reported thereon.

1 Notes on the *Geographia* of Ptolemy and on the Used Versions

The work of Claudius Ptolemy's *Geographia* is a treatise published around 140 AD in Alexandria. In it the author intends to give a global overview of the Ekumen, that is to say, the stretch of the Earth then known and inhabited, by using the most subtle mathematical and astronomical instruments of that time.

N. Lanciano (✉) · E. Cicciarelli
Rome «La Sapienza» University, Rome, Italy
e-mail: nicoletta.lanciano@uniroma1.it

N. Lanciano · E. Cicciarelli
MCE (Movimento di Cooperazione Educativa - Educational Cooperative Movement), Rome, Italy

A. Orlando (ed.), *The Light, The Stones and The Sacred*,
Astrophysics and Space Science Proceedings 48,
DOI 10.1007/978-3-319-54487-8_4

The treatise consists of eight books, the first of which contains the calculations of the Ekumen's limits, the methods for a true representation on the plane and on the sphere, the durations of the day during the summer solstice as the characteristics of the different latitudes.[1] The other books, instead, contain the geographical coordinates of the main points of interest, and probably the last book included the maps drawn by Ptolemy himself, now lost.

There are several versions of Ptolemy's *Geographia*, but in this article we decided to work primarily on the translation from Greek by Girolamo Ruscelli, drafted in 1594 in Venice. From the consultation of other translations from the Sixteenth century many differences emerged, not only of a linguistic nature: indeed, in those years the lists of the coordinates were very much used in practice, intended for consultation and the drawing of the maps, and therefore subject to constant revision.

However, it was noted that the coordinates given in the First Book are the same as in all the consulted versions, and often different from those in the subsequent books. It can therefore be assumed that the treatise has not undergone the same process of modification, but that it has remained faithful to the original ideas of Ptolemy.

We therefore decided to use as much as possible the values given directly in the First Book, and then specify when the values were extrapolated from the lists or from the maps.

2 Agreements and Differences Between Ptolemy and Marinus of Tyre

The descriptions of the Ekumen of the two geographers have many points in common, starting with the size of the planet (Ptolemy, Ruscelli's Translation 1564). Like Marinus, in fact, Ptolemy assumes that one arc of 500 stadia corresponds to one degree of arc of a maximum circle and, consequently, he argues that the Earth's circumference has a measurement equal to 180,000 stadia (Fig. 1).

Moreover, in agreement with the geographer of Tyre, Ptolemy fixes as the North limit of the Ekumen the parallel passing through the island of Thule, at 63° N, and indicates latitude 23° 50′ N for the Tropic of Cancer.

In addition, both scholars choose as zero Meridian that of the Fortunate Isles. In this paper we do not question the identification of these islands, discussed in detail in the text *L'America dimenticata* by L. Russo. Therefore, both divide the extension from West to East in hourly ranges each having 15° of amplitude.

Ptolemy deeply revises the data contained in the works of Marinus by establishing new Eastern and Southern limits. Specifically, while Marinus extends the

[1]In this article we discuss the calculation of the Ekumen's limits. For other aspects and a more in depth discussion see Cicciarelli (2015).

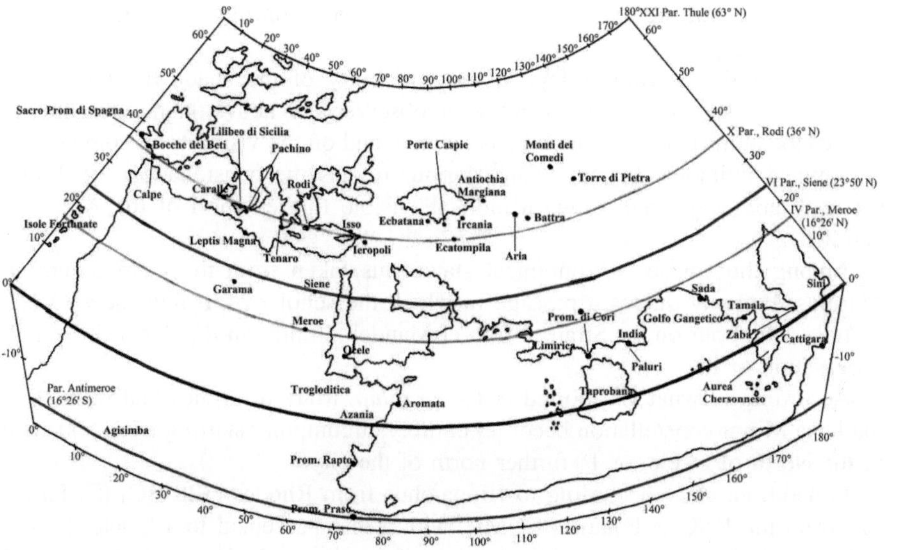

Fig. 1 Representation of Claudius Ptolemy's Ekumen

Ekumen up to latitude 24° S, namely almost up to the Tropic of Capricorn, Ptolemy does not exceed 16° 25′ S, on the parallel he calls Antimeroe and passing through the Agisimba region.

As far as the longitude is concerned, while Marinus assumes that the Ekumen consists of 15 hourly ranges for an extension of 225° E, Ptolemy considers that the ranges should be only 12, thus setting the Eastern limit at 180° E.

According to Ptolemy, Marinus would not have described territories unknown to the scholar of Alexandria, but he would have not assessed in a more appropriate way the data in his possession, resulting in excessive expansion of the known world.

3 Problems of Latitude

Marinus and Ptolemy work in an era in which the calculation of long distances is very complex and prone to major errors. Indeed, the estimates are deduced from data extracted from accounts of travels and explorations, through methods that have allowed the authors to select, manipulate and approximate the data according to yardsticks, frequently linked to personal perception rather than objective and shared criteria.

Ptolemy himself highlights the problem, arguing that the only reliable methods are based on astronomical observations. He, therefore, describes some among the astronomical phenomena cited by Marinus, with the intent of demonstrating how the latter does not provide any evidence of the extension of the Ekumen South of the Winter Tropic.

For example, Marinus refers to the variability of the shadows in the range between the two tropics to the fact that an observer sees above the horizon only the pole of the hemisphere in which he is located, and on the visibility of some stars in determined latitudes. In particular, Marinus quotes the constellations of Taurus, Orion, Canis Major and Canis Minor, all visible further north of the Antimeroe parallel.

Among the various astronomical statements taken from the Commentarii by Marinus, Ptolemy quotes a passage in which the scholar of Tyrededuces Ocele's latitude, a harbour on the Strait of Bab-el-Mandeb, in present-day Yemen, from the observation of the sky.

According to what is reported in *Geographia*, Marinus argues that in his time, the Ursa Minor constellation becomes entirely circumpolar starting from 500 stadia in the North of Ocele, or 1° further north of the city.

In addition, since according to Hipparchus from Rhodes (128 BC) the farthest star from the PNC is Polaris (α-UMi), with a distance equal to 12° and 2/5 (12° 24′), the scholar concludes that the latitude of the city of Ocele must be equal to 11° and 2/5 N (11° 24′ N).

In fact, this location's latitude is 12° 49′ N or nearly one and a half degrees farther north with respect to what was indicated by Marinus.

In the text, Ptolemy does not refute the hypothesis of Marinus, but he assigns in the lists of the *Geographia* versions, consulted in Ocele, latitude 12° N. In fact, the correction is much closer to the real value, but since the lists have been subjected to extensive and deep changes, it is not possible to attribute with certainty the coordinates to the author of the work, because it may have been modified in the following centuries.

Marinus and Ptolemy, however, do not seem to take into consideration the fact that in the second century AD, by effect of the precession,[2] the sky has changed with respect to 128 BC (Fig. 2).

From the comparison with the real declinations[3] of the Ursa Minor constellation in 128 BC, in 100 AD and in 140 AD (Table 1), we have a good correspondence with the declination assigned by Hipparchus to α-UMi. However, in the following centuries, by effect of the precession, it is Pherkard (γ-Umi) the star of Ursa Minor farthest from the PNC.

In *Almagest*, a treatise on astronomy written by Ptolemy, the scholar speaks explicitly of the precession of the equinoxes, showing to be aware of the phenomenon, while underestimating the value of the annual change of the ecliptic latitude.

This does not happen in *Geographia*, in which the phenomenon appears to be almost ignored, thereby generating the error on calculating the latitude of Ocele.

[2]See article by Elio Antonello in this volume.

[3]The declinations contained in this text were obtained through software GUIDE 9.0.

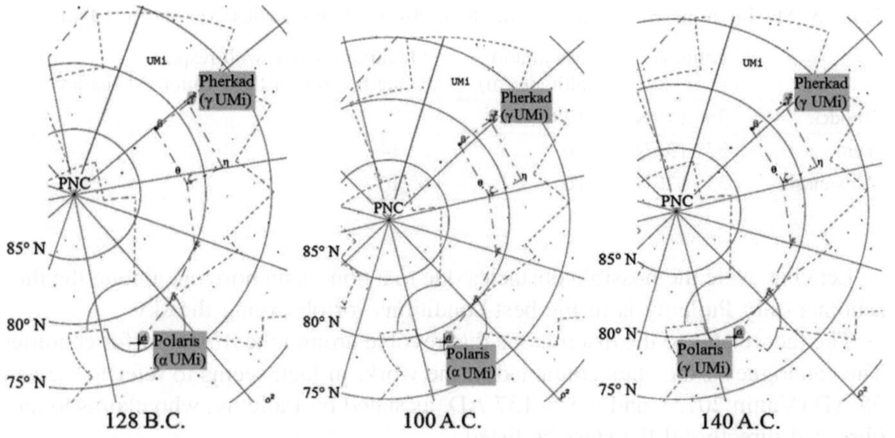

Fig. 2 The sky at the time of Hipparchus, Marinus and Ptolemy (images obtained with GUIDE 9.0)

Table 1 Declination of α-UMi and γ-UMi at the time of Hipparchus, Marinus and Ptolemy

	α-Umi		γ-Umi	
	Declination	Distance from PNC	Declination	Distance from PNC
Hipparchus (128 BC)	77° 32′ N	12° 28′	78° 53′ N	11° 7′
Marinus (100 AD)	78° 47′ N	11° 13′	78° 15′ N	11° 45′
Ptolemy (140 AC)	79° N	11°	78° 8′ N	11° 52′

Table 2 Declinations and visibility ranges of α-Carinae

Year	Declination	Visibility range
128 BC	52° 39′ S	From 37° 21′ N in the South Pole
100 AC	52° 33′ S	From 37° 27′ N in the South Pole
140 AC	52° 32′ S	From 37° 28′ N in the South Pole

Another anomaly can be found for the star Canopus (α-Car) according to Marinus visible by those who navigate from East to West along the coasts of India. The data is actually correct, but immediately followed by the clarification of Ptolemy, who argues that the star is not visible at the latitudes of the Mediterranean.

In fact, Canopus is one of the stars that is not included either in the *Almagest* catalogue of stars or in the *Commentarii* of Hipparchus. From the declinations of the star, it was possible to calculate the ranges of visibility at the times of the three scholars (Table 2).

By comparing with the latitudes of the locations where the three scientists have worked, it can be observed that in all cases Canopus is in the visibility ranges, nevertheless always reaching a rather reduced maximum with respect to the horizon (Table 3).

Table 3 Maximum height of Canopus on the horizons of Hipparchus, Marinus and Ptolemy

	Latitude	Maximum altitude (m)	Maximum height with respect to the horizon of the location (at sea level)
Rhodes	36° 14′ N	1280	1° 7′
Tyre	35° 13′ N	10	2° 14′
Alexandria	31° 12′ N	10	6° 16′

Leaving aside the possible obstacles due to a non-clean horizon, among the three scholars only Ptolemy is in the best conditions for observing the sky.

The fact relates to the discussions that revolve around the *Almagest* for centuries. The catalogue of the stars contained in the work, in fact, seems to refer to the year 58 AD (Vanin 2013), and not to 137 AD, as stated by Ptolemy, who claims to have observed directly all the stars he listed.

From the comparison with the work of Hipparchus, there emerges a compatibility between the data of the latter and the annual variation in the ecliptic longitudes taken by Ptolemy, equal to 36″ per year. This suggests that Ptolemy has simply updated the list of the Hipparchus' coordinates with the data of the precession, an idea also strengthened by the fact that in the *Almagest* only the stars visible to Rhodes' latitude are cited. But, there is no irrefutable evidence to confirm this hypothesis.

Since in 128 BC Canopus is extremely low on the horizon of Rhodes, and consequently observed with great difficulty in those latitudes. We could suppose that the star belongs to the group of stars visible in Alexandria and not named by Ptolemy because it was not present on the lists of Hipparchus. But it must not be ignored that the height on the horizon of Alexandria is not much greater than that of the horizon in the Greek island, and that therefore Ptolemy may have had the same problems as his predecessor. In addition, the coast of the Egyptian city, namely the horizon likely much clearer, extends in a north-easter direction, while in 140 AD the star culminates in a southerly direction.

Although Ptolemy prefers the analysis of the astronomical data, he does not have at his disposal enough data to obtain a good description of the Ekumen. He is therefore forced to re-work the travel data reports, both by sea and by land. The starting point consists of what is present in the works of Marinus, which Ptolemy reviews and criticises meticulously.

By relying on what had been collected during a land exploration from Leptis Magna[4] until Agisimba, Marinus places the latter at 24,680 stadia south of the Equator. Similarly, from what he collected during a sea voyage from Ptolemais in Troglodyte up to the Praso Promontory, along the coasts of present-day Somalia, Kenya and Tanzania, Marinus establishes that the promontory is at 27,800 stadia south of the Equator (Fig. 1). By using the estimate of 500 stadia per one degree of

[4]A harbour in the Mediterranean, near Tripoli, Libya.

a maximum circle, the distances correspond respectively to the latitudes 49° 22′ S and 55° 36′ S.

However Marinus is convinced that the sphericity of the Earth produces a perfect symmetry of the climates with respect to the equator, and as a result, locations placed on opposite parallels should be characterised by the same temperatures, flora and fauna.

This hypothesis does not find confirmation in travellers' descriptions of the southern regions of the world in terms of lands inhabited by exotic animals, such as elephants and rhinos, which certainly are not present north of parallel 50° N, the latitude of the Black Sea.

Ptolemy then accuses Marinus of deliberately having shortened the distance taken into consideration so that the southern border coincides with the Tropic Winter.

Subsequently the Alexandrian claims that, starting from an incorrect interpretation of the Roman expedition reports of the first century AD conducted by Septimius Flaccus and Julius Maternus, in the Commentarii a significant expansion would have been made of the distance between Leptis Magna and Garama (present-day Germa in Libya), capital of the Garamantes, an ancient nomadic people of North Africa. Moreover, according to Marinus, the parallel further to the south would pass through the Praso Promontory and the Agisimba region would be further north.

The opinion of Ptolemy is that Marinus relies too much on the reports of Flaccus and Maternus, in which it is related improperly of travels southwards, without considering the inevitable deviations. Such superficiality and the mistaken idea of the perfectly symmetrical distribution of the populations with respect to the Equator led the scholar of Tyre to overestimate the distance along the Meridian between Leptis Magna and Garama.

Although not specifying the source and method, Ptolemy quantifies the length of this stretch in 5,400 stadia, which, divided by the measurement of the arc of 1°, equal to 500 stadia, we can calculate in 10° 48′ on the arc of the Meridian, according to the lists of the versions of Ruscelli, in which the city of Garama has a latitude of 21° 30′ N, Leptis Magna 31° 40′ N, with a distance on the Meridian of 10° 10′.

The difference in latitude of the archaeological sites of the two cities is 6° 5′, not very close to Ptolemy's value. If, moreover, Ptolemy's measurements were to be converted by using the measurement of 700 stadia per 1° calculated by Eratosthenes starting from 5,400 stadia, a figure closer to the real one would be obtained, equal to a difference of 7° 42′ of latitude.

To definitively prove Marinus' mistake, Ptolemy shifts his attention to the sea voyages conducted along the coast of the Horn of Africa by such navigators as Diogenes and Theophilus. The routes of these two sailors are very similar, but while the first sails from north southwards, the second sails in the opposite direction, along a stretch slightly longer than the previous, and, according to Marinus, both travelled with a tailwind. Calculating the distance in stadia, the geographer takes for granted the value of 1,000 stadia established by Theophilus for the stretch between

Aromata (i.e. the Somali Peninsula) to the Rapto[5] Promontory, which corresponds to only 2° on the arc of the Meridian. Ptolemy intervenes arguing that it is not only the calculation of distances to be wrong, but the very distribution of the locations. In fact, he believes that the Rapto Promontory is north of the mouth of river Rapto, and not vice versa. The coordinates contained in the *Geographia* lists, nevertheless, confirm what was stated by Marinus, a sign of contradiction in the work of Ptolemy or a correction made on the work in retrospect.

Ptolemy, considering unreliable the reports of extremely long journeys, divides up Diogenes' journey into shorter journeys, showing lap by lap the days of sailing. With this calculation, he establishes that to sail from Aromata to the Rapto Promontory are needed 19 days of navigation, with an average speed of 300 or 400 stadia per day, with a distance on the Meridian of 12° 40′, with a little value lower than the real 14° 50′.

Therefore, it seems that, while Marinus works on an excessive contraction, Ptolemy has given a good estimate of the distance, but shifting nevertheless the whole strip a few degrees further south with respect to the real.

The underestimation of Marinus is partly offset by an expansion of the distance between the Rapto Promontory and the Praso Promontory, for which he accepts the stretch of 5,000 stadia of Dioscorus, against the 4,000 stadia (equal to 8°), derived from the latitudes of Ptolemy.

In conclusion, Ptolemy thinks that Marinus, to avoid contradicting that hypothesis of symmetry that caused him to make an excessive expansion of the Ekumen north of Aromata, was forced to drastically reduce the strip up to the Praso Promontory, so that the southern limit was exactly the Winter Tropic.

4 Problems of Longitude

To establish the eastern boundary of the Ekumen, Marinus and Ptolemy can rely only on data from travel reports. The Alexandrian mathematician mentions the astronomical method based on the observation of lunar eclipses, but states that he does not have a sufficient amount of data for a complete discussion. He then decides to take the information contained in the *Commentarii*, to process and correct them.

Specifically, the Ekumen is divided into three courses, the first two by land, of which the first from the Fortunate Isles to the Euphrates River pass near Ieropoli (171° 15′ E), the second from the Euphrates to the Tower of Stone (135° E), and the last by sea from the Cori Promontory (125° 20′ E) to Kattigara (177° E).

Marinus' calculation for the distance between the Fortunate Isles and the Euphrates River pass, near Hierapolis is accepted and confirmed by Ptolemy.

The Alexandrian indeed observes that the journey, divided into shorter laps, develops indicatively along the Rhodes parallel (36° N), and that the measurements

[5]River Rapto today is known as Galano or Sabaki, and is the second river in Kenya.

in stadia are converted into degrees with the right proportion, or considering that a measurement of 400 stadia corresponds to one arc of a degree of the parallel.

As a matter of fact, the two work an expansion with respect to the truth, bringing Hierapolis about 16° further eastwards.

The situation is different for the next stretch, which is also divided into laps, but this time it does not develop along the same parallel. According to Ptolemy, Marinus does not take this element into consideration, and it is therefore necessary to reduce the measurement contained in the *Commentarii*, setting the longitude of the Tower of Stone 6° further westward than that of Marinus.

Without the certainty of the actual position of the site, a detailed comparison between the Ptolemaic coordinates and the actual position is not possible. Assuming the longitude range suggested by archaeologists (Giulmia-Mair et al. 2009) is up to date, it is likely that it was located in today's Kyrgyzstan, with a longitude with respect to that of the Greenwich Meridian between 87° E and 92° E. It therefore follows that Ptolemy moves the Tower by about 40° to 45° eastward with respect to its actual position.

For the last stretch, once again divided into laps, Ptolemy starts from the data on stadia reported by Marinus, by subtracting one third of the value, and then chooses from time to time whether to apply a further correction and if so which. The criteria adopted are not adequately explained, but we can assume that Ptolemy has exploited the latitude differences of the various localities. The following Table 4 shows the data related to each stretch, with the coordinates extracted from the lists of *Geographia*.

In *Almagest* Ptolemy extracts the final formula of the spherical rectangular triangles from Menelao's Theorem:

Theorem (Final formula of the spherical rectangular triangles) *Let ABC be a right spherical rectangular triangle in C; let a and b be legs, and c the hypotenuse. Let* α *be the angle in A, the relationship* $\sin a = \sin \alpha \cdot \sin c$ *is valid.*

By applying the result and comparing it with the ratios used by Ptolemy we obtain the data showed in Table 5.

It thus appears that, in four cases out of six, $\sin \alpha$ is very close to the correction ratio used while in one of the stretches a comparison with $\sin \beta$ is obtained.

Table 4 Differences in longitude, latitude and geodesic of the various laps

Stretch	Difference of longitude (a)	Difference of latitude (b)	Geodesic distance (c)
Cori—Paluri	11°	7°	12.74°
Paluri—Sada	18°	4° 20′	18.05°
Sada—Tamala	3° 10′	6° 20′	7.05°
Tamala—Aurea Chersonneso	7° 30′	10°	12.48°
Aurea Cher-sonneso—Zaba	5°	2°	5.38°
Zaba—Kattigara	8° 40′	13°	15.82°

Table 5 Differences in longitude, latitude and geodesic of the various laps

Strech	sin α = sin a/sin c	sin β = sin b/sin c	Ratio by Tolomeo
Cori—Paluri	0.865	0.553	5/6 = 0.833
Paluri—Sada	0.997	0.244	1
Sada—Tamala	0.450	0.899	5/6 = 0.833
Tamala—Aurea Chersonneso	0.604	0.803	2/3 = 0.667
Aurea Cher-sonneso—Zaba	0.929	0.372	1
Zaba—Kattigara	0.553	0.825	2/3 = 0.667

In the latter case, instead, there is a significant difference in both reports. But it should be noted that is curious that the result obtained by Ptolemy for the extension in longitude equals exactly half of the Earth's circumference. The suggestion then is that the scholar, not being aware of the measurement in stadia for the last two stretches, has fixed for this purpose the length of the last itinerary with the intent of obtaining for Kattigara a longitude of 177° E, and therefore locating the extreme City of the Sini exactly at 180° E.

Therefore, from an estimation starting from the differences in latitude, Ptolemy could have chosen a correction method that meets the following guidelines:

- $b \leq 4° 20'$: no correction applied;
- $6° 20' \leq b \leq 4° 20'$: reduction of 1/6 on the total value;
- $b \geq 12°$: reduction of 1/3 on the total value.

5 The Great Meridian Line of St. Nicholas in the Benedictine Monastery in Arenis, Catania

The Meridian line is drawn on lava slabs with zodiac figures drawn by Albert Thorwaldsen on red marble. Abbot Federico La Valle had the first idea of tracing a Meridian Line in the Benedictine complex; Abbot Ansalone later assigned the project to Nicolò Cacciatore, Director of the Palermo Observatory, who in 1837 opened the gnomonic hole and prepared the base. Father Corvaja later hired the two scientists who were at that time in Sicily, Sartorius von Walterhausen of Goettingen and the Dane Peters in 1841, for the final completion (Lanciano and Celi 1998; Trobia 1999; Tuscano 1999).

There are several errors, both in the text of the literary *Cabinet Journal of the GIOENIA Academy in Catania—1840—Sciuto Brothers Typography,* in the text entitled *"On the Meridian NEWLY BUILT IN THE Church of the Benedictines in Catania—2nd Letter by N.N. to Mr. N.N.* and also in the last restoration. We read: *"A' due lati di questa linea di centro, alla uguale distanza di once undici, ossia 250 mm trovansi due altre strisce nere della stessa pietra paragone, o sopra di questa, nelle loro rispettive distanze, veggonsi segnati i numeri dei giorni di ogni mese, i quali tanto più vanno tra di loro avvicinandosi, quanto più si approssimano al solstizio di està."*

In the journal we can read: *"Lo spettro solare vi passa con un diametro il maggiore in inverno di 938 mm, ed il minore in està di 228 mm senza la penombra; ed il suo passaggio è così sensibile, che il momento puro astronomico del meridiano può facilmente sorprendersi dall'attento osservatore; imperciocché l'altezza dello gnomone essendo sopra la linea della meridiana di palmi 92,7 lo spettro dacché comincia col suo disco a passar sulla linea, finché vi giunge col centro, scorrono molti secondi, e l'occhio può ben accompagnarlo nel corso."*

Fig. 3 The passage of the light spot 19th April 2015 (photo by authors)

Fig. 4 The passage of the light spot 11th September 2015 (photo by authors)

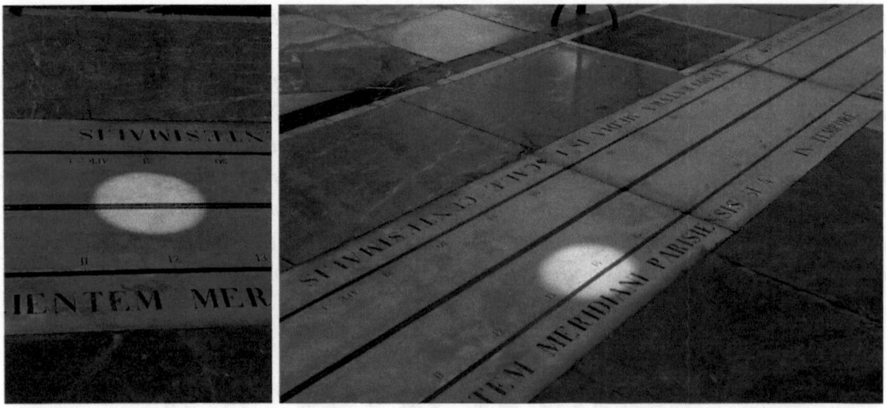

Fig. 5 The passage of the light spot 12th September 2015 and 14th September 2015 (photo by authors)

The height of the hole, open on the white vault of the chapel on the left of the main altar, is 23.92 m; while the length of the Meridian Line is 37.7 m and with inscriptions at the ends 40.92 m. It has 365 daily slots, as they were marked on the Augustus Meridian Line in Rome.

Along the line there are several data of astronomical and geographical nature: we highlight those with a special connection to what is indicated by Ptolemy as characterising the latitudes and the longitudes of the locations, and to those relating to the shape of the Earth (Figs. 3, 4 and 5).

6 Black Inscriptions in Latin on Carrara Marble Along the Whole Line

The inscriptions that we can read are:

- *Obliquitas ecliptices apparens 23° 27′ 42″,41*
- *Latitudo geographica borealis 37° 30′ 15 5″*
- *Longitudo geographica ad orientem insulae Ferri 32°46′ 0″* (Longitude from Hel Hierro in Canary Islands before 1884)
- *Sive ad orientem meridiani parisiensis 51′ 4″ in tempore* (and longitude from meridian of Paris)
- *Elevatio pavimenti supra superficiem maris 38 m 530*
- *Altitudo gnomonis 23 m 917*
- *Longitudo horizontalis a gnomone ad punctum solstitii hiemalis 43 m 034* (Two related data with Latitude)
- *Diei longitudo maxima 14 h 42,3 Minima 9 h 31,0* (Maximum and Minimum Duration of the day related to Latitude)

- *Crepusculi Astronomici Maxima Jun. 21, 1 h 59,2 et Dec. 21, 1 h 38 8, Minima Mart. 7 et Oct. 8, 1 h 31 0*
- *Crepusculi Civilis Maxima Jun. 21, 0 h 39,2 et Dec. 21, 0 h 36,9, Minima Mart. 16 et Sept. 28, 0 h 32,8* (Length of civil twilight, Sun 6° below the horizon)
- *Poli depressio l:300,7* (Shape of the Earth)
- *Terrae Semiaxis 6355949,6 m*
- *Longitudo penduli simplicis per minutum secundum vibrantis 992 mm 89* (to know the local gravity)
- *Constans gravitatis 9,799*
- *Intensitas absoluta vis magneticae terrestris horizontalis 2.4505 ad unitates millimetri, milligrammatis, et minuti secundi temporis medii revocata*
- *Declinatio acus megneticae occidentalis 15° 17′ 51″,1*
- *Inclinatio acus megneticae 54° 14′ 45″*
- *Status medius barometri ad mare relatus 762 mm 32*
- *Temperatura media 18° 1 scalae centesimalis*
- *Longitudo unius gradus meridiani 110972 m 74, paralleli 88408 m 71*
- *Altitudo Montis Aetnae 3303 m 7, Turris philisophi 2924 m 2, Monasterii S. Nicolai de Arenis 857 m 9*

At one end we read: *"misure lineari più usate all'epoca negli scambi commerciali:*

1 piede parigino è rappresentato da 3 sezioni × 35.567 cm = 97.7 cm
1 piede inglese da 3 × 30.48 cm = 91.44 cm
1 palmo siciliano del 1809 da 4 × 25.81 cm = 103.24 cm
1 palmo napoletano del 1840 da 4 × 26.365 cm = 105.46 cm
French meter of 100 cm

At the other end it reads with characters very spoiled the names of the authors and data of the meridian line: (Fig. 6)

INVERNERVNT AC R _____N_ INSIGNS ASTRONOMI
WOLFGANG SARTORIVS
DYNASTES WALTERSHAVSEN A GOTTINGA ET
DOCTOR CHRISTIANVS FRIDERICVS PETERS
A FLENSBVRGO IN D__A (DANIMARCA?)
QVI VSI SVNT OPERA CATANENSIS SCVLPTORIS
CARLO CALÌ
AVSPICAVIT D. IOHAN _____CO CORVAJA
HVJVS __ONATE__ RATE
__NSENNA_O
ANNO AERA ___VCCALIS MDCCCXLI

Fig. 6 Figures of the meridian line

7 The Other Internal Meridian Lines in Sicily

In the Metropolitan Cathedral of the Blessed Virgin Mary of the Assumption in Palermo, there is the simple Meridian line by astronomer Giuseppe Piazzi of 1795, without inscriptions but with beautiful marble inlays of the 12 signs of the zodiac. The line crosses the aisle for 21.812 m and the hole is 11.776 m above ground.

In the Protometropolitan Cathedral of the Blessed Virgin Mary of the Assumption in Messina, a Meridian line with different indications was traced between 1802 and 1804 by mathematician Antonio Maria Jaci. The degrees of the ecliptic in relation to the 12 signs of the zodiac, the hours and the minutes of the *Italian hours*, the diameter of the light spot, the declinations of the Sun and its heights above the horizon. The Meridian was first damaged in the earthquake of 1908 and disappeared completely following bombing in 1943. There remains the bell tower with automata, the perpetual calendar and the planetary, by the Ungerer brothers of Strasburg.

In the Cathedral of the Most Blessed Mary our Lady of the Annunciation in Acireale, a Meridian Line was traced in 1843 by the Danish astronomer F. Peters. The black line is surrounded by a red curve, the measurement of the transverse axis of the ellipse of the sunspot, which runs from 9 cm towards the SE to 18.2 cm towards the SI. The gnomonic hole is 9.104 m from the horizontal plane and the line is 16.452 m long: just like that of Catania, many data of astronomical and geographical character are marked.

In the Mother Church of the Blessed Virgin Mary of the Assumption in Castroreale (Messina), in 1854, a Meridian Line was traced by N. Perroni Basquez who was neither an astronomer nor a mathematician, but a professor of ancient literature and a lover of astronomy. The line has suffered from the earthquakes of 1894, 1908 and 1978 and is engraved on the floor in coloured parallelograms. It indicates, with numeric values the hours on one side and the minutes on the other side, the ancient hour of noon by the bell tower for every day of the year. It is 958 cm long and the names of the 12 months are inscribed. The gnomonic hole was at a height of 958 cm, but the plate with which it was made came off the wall and was lost.

In the Holy Mother Church of The Most Holy Apostles Peter and Paul in Castiglione di Sicilia, at the foot of Etna, there is a line of 1882 by Temistocle Zona with an 8-curve with the names of the months.

In the Mother Church of St. George in Modica, the mathematician Armando Perini traced an 8-curve Meridian Line in 1895, therefore for the true time and the mean time. It is 22 m long and is traced on white marble. The *Italian hours* can be read thanks to the passage of the Sun in the Meridian Line and the dates of entrance into the 12 signs of the zodiac indicated with their names. The hole is located at 14.18 m from the ground.

In the Royal Technical School in Caltanissetta, an ancient monastery of the Minimi Fathers, in 1913 a Meridian Line had been traced on the floor of the Management's Office by Baron Mario Bonfiglio, a scholar of gnomonic: there were the line of the local noon with the correction of 3′ 44″ in relation whit the meridian 15° Est, and also the two lines of 11.30 and 12.30 local time, in addition to the lemniscate of the mean time, on which were marked the days 4 by 4. The signs of the zodiac were drawn in 12 circles at a stretch of daylight hyperbolas, in the days of the entrance of the Sun in the signs. Of this Meridian Line there remains only a drawing by G. Musotto in a Popular Astronomy essay of 1914.

References

Cicciarelli, E.: La Geografia di Tolomeo e le critiche a Marino di Tiro, thesis of master's degree in Mathematics, supervisor N. Sapienza University, Rome, Lanciano (2015)

Giulmia-Mair, A., Jeandin, M., Ota, K.: Metal trade between Europe and Asia in classical antiquity, in metallurgy and civilisation: eurasia and beyond archetype. J. Mei e T. Rehren (eds.), London (2009)

Lanciano, N., Celi, V.: Le meridiane interne in Sicilia. Astronomia UAI **3**, 25–32 (1998)

Ptolemy, C.: Geografia, Translation from Greek by Girolamo Ruscelli, Venice (1564)

Trobia, M.: La meridiana di Castroreale, Atti IX Seminario Nazionale di Gnomonica, pp. 367–369 (1999)

Tuscano, M.L.: Meridiane e blasoni: la meridiana di Caltanissetta, Atti IX Seminario Nazionale di Gnomonica, pp. 371–372 (1999)

Vanin, G.: Catasterismi, Edizioni Rheticus-DBS Zanetti, Feltre (Belluno) (2013)

Archaeoastronomical Analysis of the Temple of Diana to Cefalù (Sicily)

Andrea Orlando and Davide Gori

Abstract The so-called *Temple of Diana* is situated on the Rock of Cefalù, that dominates the namesake fishing village in the northern coast of Sicily. The megalithic temple has a main entrance direct to the West, through which starts a corridor leading to the rocky cistern characterized by a dolmenic coverage. After the first drawings and reliefs between '700 and '800, respectively of Jean Houel and George Nott, the first official archaeological excavations were made by Pirro Marconi in the first half of the twentieth century. These excavations allowed to acquire more informations about the age of the temple and the cistern. With this study we present the first complete archaeoastronomical analysis of the building, allowing to find out that the megalithic architecture is a real Sun temple. The front door of the temple is indeed oriented to the point where the Sun sets at the equinoxes. This finding suggests that in these periods of the year the solar hierophany most likely invited to come inside the temple to reach the cistern, where they carried cults and rituals related to water. This study propone to identify the temple as an Artemision. The study made it possible to realize the first 3D relief of the temple also, suggesting a new process for the protection and enhancement of the archaeological site.

A. Orlando (✉)
Catania Astrophysical Observatory (INAF), Catania, Italy
e-mail: aorlando@oact.inaf.it

A. Orlando
Southern National Laboratories (INFN), Catania, Italy

A. Orlando · D. Gori
Institute of Sicilian Archaeoastronomy (IAS), Novara di Sicilia, Italy

D. Gori
Amec Foster Wheeler Italiana, Corsico, Italy

A. Orlando (ed.), *The Light, The Stones and The Sacred*,
Astrophysics and Space Science Proceedings 48,
DOI 10.1007/978-3-319-54487-8_5

1 The Rock of Cefalù: Tectonic and Geological Framework

The *Rock of Cefalù* is a large limestone plateau that rises about 300 m on the Cefalù's promontory, situated on the Tyrrhenian coast, to the eastern borders of the province of Palermo, in northern Sicily (Fig. 1). The rocks that outcrop belong to Panormide Tectonic Unit (Unit Maghrebids "external") being a sequence of Mesozoic carbonate platform with a siliciclastic and marly tertiary coverage.

In particular there is the formation of Cefalù (CEU), Upper Jurassic—Early Cretaceous age, characterized by reef carbonate limestone gray-bluish sometimes pseudoolithics (Fig. 2); there are levels of reddish or yellow-ocher intraformational breccias also. The limestones are fossiliferous and contain rudist, gastropods, corals, algae and bryozoans.

Structurally and tectonically this area is characterized by the presence of dislocation lines (normal faults) mainly in the northeastern area of the "Rocca" (Lunardi et al. 1994).

The micro-karst geomorphology it's very interesting; the limestone is dissected into blocks, called clints, bounded by vertical fissures known as karst crevasses (*Kluftkarren*). Apart from clints and crevasses, limestones also have a number of characteristic surface formations known as runnels (*Rillenkarren*), furrows (*Rinnenkarren*), dissolution/corrosion pans and tubs.

Fig. 1 The Rock of Cefalù and the town photographed from East (authors' photo)

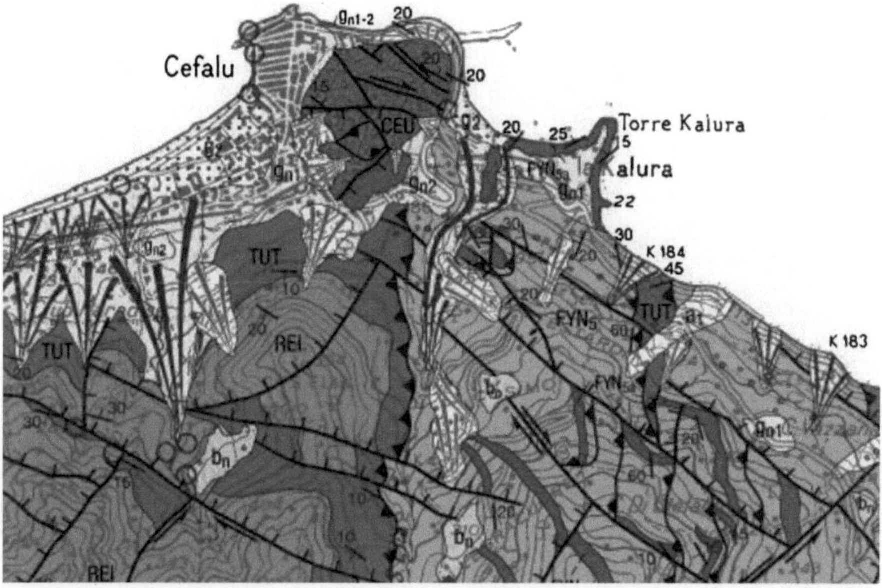

Fig. 2 Geological map of "Rocca di Cefalù" (modified from Carta Geologica d'Italia 1:50.000—Foglio 597 Cefalù, ISPRA)

2 Temple of Diana: Description

There are some reliefs and descriptions about the megalithic temple and the dolmenic tank in the works of Houel (1785), Hittorf and Zanth (1829) and Nott (1831).

One of the first representations of the megalithic temple of Diana is that of the famous French painter and architect Jean Houel, one of the greatest travelers of the Grand Tour. During his trip to Sicily Houel realized over 200 drawings, which will be collected in the four volumes of the *Voyage pittoresque des isles de Sicile, Malta et de Lipari* (1785). This collection is considered one of the most important works of the eighteenth century. From the drawing of the temple on the Cefalù Rock (Fig. 3) we note in particular the three rows of stones above the main door.

The first reliefs of the temple of Diana was performed by George Nott, an Anglican clergyman, scholar and lover of Italian culture, that lived in between '700 and '800.

The general plan of the building (Fig. 4) shows the original megalithic structure (dark lines) and adding medieval church (gray lines). We see three doors from the plant (H, L, J), the corridor (I), the main chamber (M) and the smaller secondary room (K). With the letter Y Nott indicates the cistern carved into the rock. Reliefs of the elevation of the west facade (Fig. 5) is to emphasize the presence of the gargoyle gutter that was used to drain the water from the terrace or roof that covered the large room. Even from Nott's reliefs it can still be seen three rows of stones above the front door.

Fig. 3 The temple of Diana in the Jean Houel's drawing (1785)

Today the most evident part of the temple's ruins consists of a building leaning against the rocky slope of the mountain, in the western part of the Cefalù Rock, where is the only route access to the structures present on it. The floor plan of the temple is irregularly rectangular: the longer side of the building, facing west, is not continuous, in fact in its southern part it has a greater rectum stretch (11.45 m), while proceeding towards the north it pronounces a recess, and continues for 7.75 m, backward of 2.10 m from the remaining front. The temple is preserved to a considerable height, and therefore constitutes a ruin of important dimensions; on the South-West corner the greater height amounts at 5.07 m.

Also the corners of the temple, except the north one's, have a rudimentary decorative work in a kind of pilaster.

The front door is not at the center of the building, but near the North-West corner. The door is composed of a monolithic lintel, 2.58 m long and 0.69 m high, placed on two jambs of 0.65 m wide, composed of various segments, connected to the walls of the building. The door is 2.68 m high and has a light of 1.19 m at the bottom and 1.16 m on the top. Entering through the door starts the corridor, approximately 7 m long, which leads to the dolmenic cistern, whose coverage is set to a higher share of about 4–5 m above the megalithic temple's floor. Each of the walls that surround the corridor has a door that leads into the two side compartments of the building (the north room has dimensions of 3.05 × 5.00 m, while the south room 7.85 × 4.14 m).

In the largest room was built a church dedicated to Santa Venera in high medieval age (Brunazzi 1997).

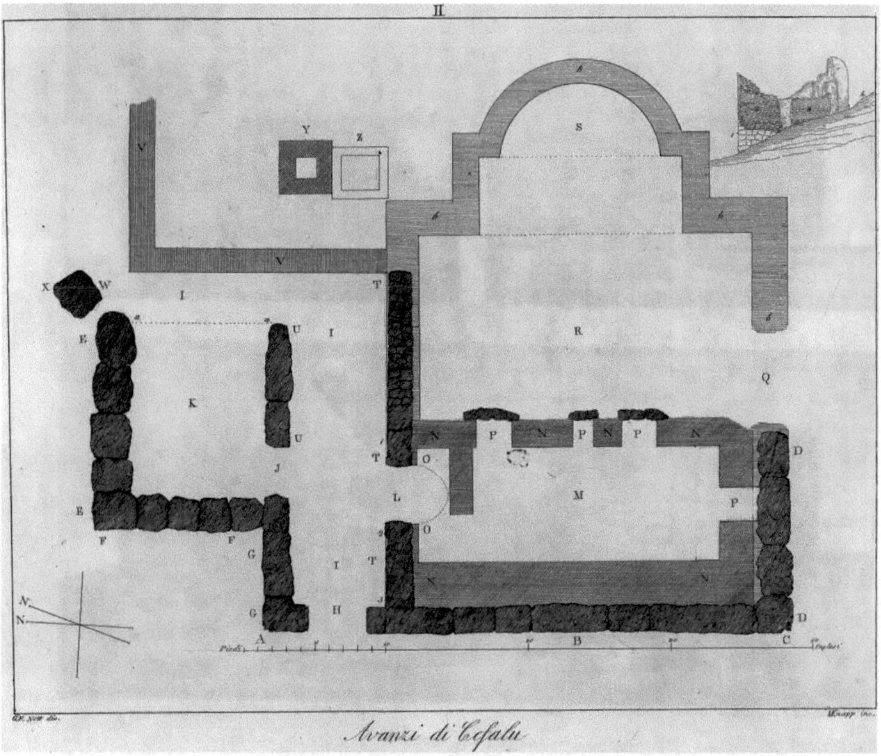

Fig. 4 The Temple of Diana's plan (from Nott 1831)

While the front and sidewalls are remarkable elements, as regards the eastern rearward part the elements are still very poor. All the walls are formed of blocks placed in situ, as they had been plucked from the mountain, and they were assembled to dry, in fact the mortar present in some part of them is of recent times. The size and shape of the blocks are very different, there are colossal stones and other rather small. These blocks should be to form three rows of blocks in height (Fig. 5). From different photos taken from the beginning of XX century until today it is evident that the three rows of blocks are no longer present, there are in fact only small portions of the rows of stones (Figs. 6 and 7).

2.1 The Dolmenic Cistern

The cistern is located on a deep pit of the rock, a veritable little sinkhole, which has a small pool of water used since ancient times as a reservoir of water. The tank has a surface area of 19 m^2 with an elliptic shape.

Fig. 5 Reliefs (from the *bottom* to the *top*) of the elevation of the west façade (*III*), the access door to the small room (*IV*), the outer wall of the small chamber (*V*), the access door to the large room (*VI*), particular of the gargoyle (*VII, VIII*), moldings (*IX*) (Nott 1831)

Fig. 6 The Temple of Diana in a rare photo of the beginning of the '20s of the twentieth century, before the Pirro Marconi's excavations (courtesy of Photo-Archive and Photo Library by Varzì and Brunetti, Cefalù)

The "primordial" tank is a little cavity in the limestone bedrock, caused by the wellspring water erosion until a corrosion tub. The cistern is located on a deep pit of the rock, a veritable sinkhole, which has a small pool of water used since ancient times as a reservoir of water. The tank has a surface area of 19 m^2 with an elliptic shape (Fig. 8).

The maximum depth of the tank is about 4 m, while the bottom, flat, occupies an area of just 6 m^2. The tank was then closed with a dolmenic type cover. A series of large slabs of limestone in fact rest on the edges of the cavity and on two rectangular monoliths that act as lintels, which in turn are imposed on a central pillar formed by 4 stones suitably machined and overlapped, three of which are of cylindrical form while one is rectangular. The pier is located at the center of the cavity and thus sustains the entire dolmenic roof, formed by seven irregular slabs: 4 monoliths depart from the midline (lintels) to the west, while the others 3 are directed from the opposite side to the east. The greatest segments measured 2.60 m in length, others arrive at 2.30 or 2.40 m, while the width varies between 1÷1.75 m. The heads of the slabs rest directly on the edge of the rocky cavities, without any recess housing had been practiced to accommodate them (Fig. 9).

Fig. 7 The Temple of Diana as shown today (authors' photo)

2.2 *The First Archaeological Excavations*

The first area's archaeological excavations on the so-called Temple of Diana were conducted by Pirro Marconi in the first half of the twentieth century. The archaeological investigations, regarding the temple structure, led to the discovery pottery fragments dating from Greek times, as painted shards, fragments of tiles and pithoi, and from Middle ages, like fragments of Norman dishes and Byzantine pottery. As regards the dolmenic cistern were found several fragments of impasto vases, reddish and yellowish, recalling without any doubt to prehistoric times, probably to the period of the advanced Bronze Age, at the beginning of the first millennium BC (Marconi 1929).

So the megalithic temple seems to date back to the VI–V century BC while the dolmenic cistern incorporated on it is considered of proto-historic period by many scholars (Bovio Marconi 1956; Van Essen 1957; Tusa 1959; Tullio 1974). It is interesting at this time to emphasize how Marconi already proposed a parallel between the function of the megalithic complex present on the Rock of Cefalu and that of Agrigento, the archaic rocky sanctuary situated on the Rupe Atenea.

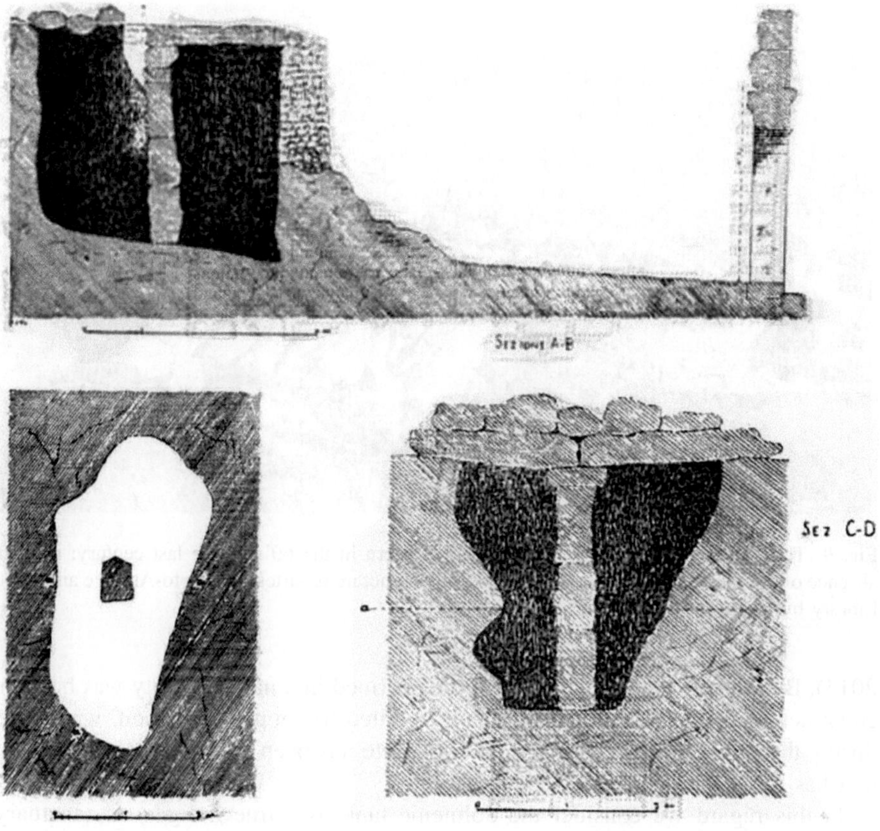

Fig. 8 The dolmenic cistern (from *top* to *bottom*, *clockwise*): North-South section, East-West section, plant (from Marconi 1929)

3 The Cult of Water

The study of the Temple of Diana and the dolmenic tank is really very interesting, as it still is not well understood what was their real function. As already highlighted by Marconi seems certain, however, that the larger building is closely linked to the tank: in fact the axis of the building, passing through the main door and the hall tip directly on the tank.

Thereby the temple acts as pronaos: to access on the tank you have to go through the building. So the function of the complex not can be that sacred. The worship of water is among the most frequent and widespread in prehistoric times, in all regions, and in certain (remember Sardinia, for example: Melis 2008) assumes the demonstrations also of great architectural importance (Portale 2012; Collin Bouffier

Fig. 9 The temple of Diana with the dolmenic cistern in the 60's of the last century; note the absence of the pine forest in front of the megalithic structure (courtesy of Photo-Archive and Photo Library by Varzì and Brunetti, Cefalù)

2013). By the cult of a pool of water being formed in a natural cavity was born the construction of the sacred place, initially limited to simple protection, while later during the Greek period, access to the holy site has been barred from the building that has the value of pronaos and service.

In this regard we consider the dolmenic tank as a true indigenous sanctuary, while the temple would be identified with an Artemision. It is to remember also that in the Middle Ages on the ruins of the Temple of Diana, in particular on the area to the right of the main entrance, it was built a church dedicated to Santa Venera. It is interesting at this point to consider other possible sites and facilities related to the water's cult in Sicily. The first of these is the Temple of Diana Facellina in Milazzo. Today there are no more traces of this temple, however it has identified a thermal building in the land of Santa Venera near Milazzo (Saporetti 2008).

In this regard we quote an interesting passage from his book: "*We are likely to see an origin of the sacredness of the place in a rich source of healing water, perhaps even worshiped before coming of the Greeks and the consecration of the goddess Artemis/Diana*".

We must also emphasize the fact that the presence of a church dedicated to Santa Venera, built on a sacred pagan site, is not meaningless and not a unique case. In the province of Catania in fact, in the territory of Acireale, we find the archaeological area of *Santa Venera al Pozzo* (Fig. 10).

The presence at the place of enormous quantities of water has led over the centuries the location of structures that could contribute to its exploitation: the baths, the well, the mills, the ducts, the hospital and the buildings dedicated to

Fig. 10 The archaeological area of Santa Venera al Pozzo, in the image are showed the remains of a thermal plant consisting of two rooms covered with vaulted roof used as bathrooms (authors' photo)

religion. Even long before the cult of Santa Venera, under the "Timpa", in front of the main facade of the church dedicated to the Saint, it was to be already a center of worship attested by the discovery of clay figurines related to the cult of Demeter and Kore, protective deities of the land and agriculture (Branciforti 2006; Amari 2006).

And finally we must certainly remember the rocky sanctuary in Agrigento, a place that already from Marconi was identified as a sanctuary of the period before the Greek colonization (Marconi 1926). A place intended for worship of natural forces, perhaps dedicated to nymphal deities (Bellavia et al. 2012; Portale 2012; Fino 2014).

The sanctuary is characterized by two natural caves from which water flows, and which is then collected through a system of tanks. Today the rocky sanctuary is under study yet and is not visible to public.

3.1 The Archaeoastronomical Analysis

Initially we measured the orientation of the megalithic temple using satellite data (tool of GE). From this verification it showed that the azimuth of the entrance of the Temple of Diana is about 270° (Fig. 11).

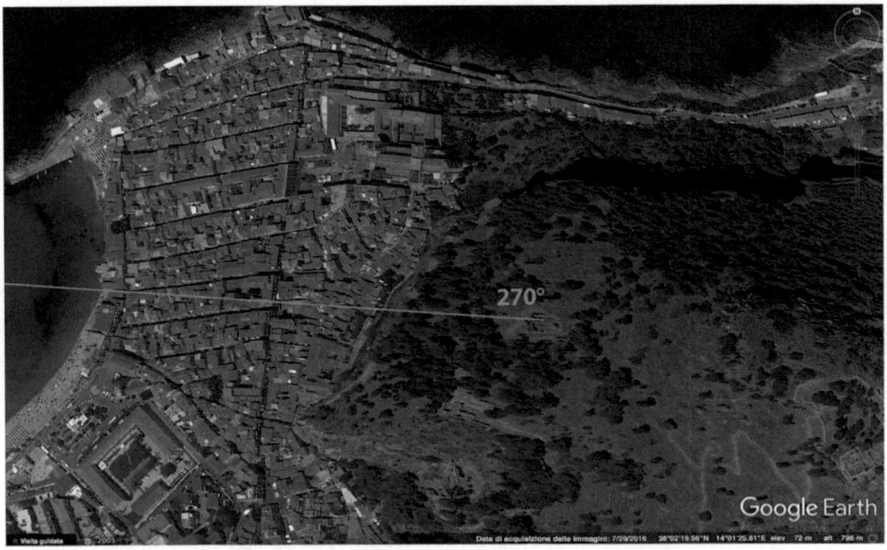

Fig. 11 Rock of Cefalù. Satellite view of Temple, with the Google Earth ruler showing the azimuths of the building's entrance (Image courtesy Google Earth, drawing by the authors)

Table 1 The table shows the azimuth from inside looking out, the angular height of the horizon in that direction, the corresponding declination and the altitude

	Azimuth (°)	Height (°)	Declination (°)	Altitude (masl)
Temple of Diana's entrance	268.50	0.23	−1.14	154

Then in August 2015 we have achieved the campaign of measures on the Rock of Cefalù. In particular, the study was carried out to measure the azimuth of the Temple of Diana's entrance. To implement the measures we used a theodolite Kern DKM2 of Swiss manufacture. Although the nominal accuracy of the instrument is less than 1', it is estimated that the error of our measurements can reasonably be ±15'. The results are shown in Table 1.

So it was found as the orientation of the front door of the megalithic temple, which leads directly to the corridor, has a clear direction: East-West. The azimuth of the entrance and the corridor of the temple is clearly equinoctial, and this indicates perfectly the direction in which the Sun sets at the equinoxes. Unfortunately, about 50 years ago in front of the temple several pine trees have been planted, so today it is not possible to observe the western horizon and thus the equinoctial hierophany (Fig. 12).

To observe the Sun setting at the equinoxes in alignment with the front door it was realized a virtual 3D reconstruction video using the technique of architectural photogrammetry (Fig. 13).

Fig. 12 The Temple of Diana photographed from East: to notice the pine forest that unfortunately does not allow to observe the Sun sets in alignment with the entrance of the temple at the equinoxes (authors' photo)

Fig. 13 Temple of Diana, entrance: view from inside towards the outside. *Left* photo of 1963 when it was still possible to observe the sea and the western horizon (courtesy of Photo-Archive and Photo Library by Varzì and Brunetti, Cefalù); *right* 3D video reconstruction that shows the moment when the Sun sets at the equinoxes

Fig. 14 *Left* and *right*: two images extract from the 3D Diana's Temple render

3.2 Diana's Temple 3D Model

In order to obtain a georeferenced (UTM-WGS84) 3D model of the megalithic structure, was executed a "non-invasive" photogrammetric survey and a take-over with "total station".

The activities were structured as follows: the megalithic area's inspection, non-invasive and non-destructive survey, post data processing, return the 3D model.

The technologies used during the temple take-over were (Lo Brutto and Spera 2011; Remondino et al. 2014): "total station" for detail points (for object scale commissioning), GPS for reading station points necessary for georeferencing the 3D model in the geographic coordinate (UTM-WGS84), Reflex Nikon camera for photogrammetric survey, targets on rigid support to facilitate the points reading, "Disto Laser" for a checking during the model return.

At the end of the different calculations were produced a 3D model, a render video of the sun sunset at the equinox, a series of 3D Diana's Temple render images and a 3D PDF file (Fig. 14).

A digital copy of the final work was donated to the Municipality of Cefalù to begin a process of protection and enhancement of the Diana's Temple area creating the first "interactive archaeoastronomical museum" of Sicily.

4 Conclusions

The Temple of Diana on the Rock of Cefalù is certainly one of the most attractive monuments in Sicily, the importance of which was little considered for long time.

The archaeoastronomical study carried out made it possible to again refocus attention on it, allowing to start a new protection and enhancement process. The archaeoastronomical analysis of the temple has uncovered as its entrance is oriented to the sunset at the equinoxes. The Temple of Diana is configured thus as a real Sun temple, a place where during the sunsets at the equinoxes could be observed a magnificent hierophany: the Sun, in alignment with the front door, illuminating the hallway of the temple, inviting participants to go on until the megalithic tank, center of an atavistic water's cult. And it is the presence of water that has led us to propose a new allocation of the temple, considering the goddess Artemis; so we could consider the place as an *Artemision*.

This study was officially presented to Cefalù during a meeting organized in cooperation with the Administration. On January 9th 2016, in the council chamber of the Town Hall, a packed crowd attended the cultural event dedicated to the archaeoastronomical study of the Temple of Diana. Now we hope that in the coming years, the Administration can realize a multidisciplinary tour route and the first 'interactive archaeoastronomical museum', to be implemented most likely in one of the rooms of the *Osterio Magno*, the medieval palace located in the center of Cefalù.

References

Amari, S.: I materiali in esposizione nell'Antiquarium. Sale I-II-III, Palermo (2006)

Bellavia, M., et al.: Nuovi dati sulla Chiesa di San Biagio ad Agrigento. Archeologia Postmedievale **16**, 125–132 (2012)

Bovio Marconi, I.: I monumenti megalitici di Cefalù e l'architettura protostorica mediterranea. In: Atti del VII Congresso Nazionale di Storia dell'Architettura, Palermo 1950, Palermo, pp. 213–221 (1956)

Branciforti, M.G.: L'area archeologica di Santa Venera al Pozzo-Acium. Antiquarium, Palermo (2006)

Brunazzi, V.: La chiesetta medievale sull'edificio megalitico della Rocca di Cefalù. In: Archeologia e territorio, Palermo, pp. 361–374 (1997)

Collin Bouffier, S.: Il culto delle acque nella Sicilia greca: mito o realtà? In: Teti, V. (ed.) Storia dell'acqua. Mondi materiali e universi simbolici, Donzelli Editore, Roma (2013)

Fino, A.: La fontana arcaica di San Biagio ad Agrigento. In: Livadiotti, M., Parello, M.C. (a cura di), Il restauro dei monumenti antichi. Problemi strutturali: esperienze e prospettive, Atti delle Giornate di Studio, Agrigento, 23–24 novembre 2012, Thiasos, 3.2, Convegni, pp. 67–91 (2014)

Hittorf, J.I., Zanth, L.: Architecture antique de la Sicile. Annali dell'Instituto di Corrispondenza Archeologica, Roma, pp. 362–370 (1829)

Houel, J.: Voyage pittorosque des iles de Sicilie, IV, p. 92, tav. XLIX-LI, Parigi (1785)

Lo Brutto, M., Spera, M.G.: Image-based and range-based 3D modelling of archaeological cultural heritage: the Telamon of the Temple of Olympian Zeus in Agrigento (Italy). In: 4th ISPRS International Workshop 3D-ARCH 2011: "3D Virtual Reconstruction and Visualization of Complex Architectures", Trento, Italy (2011)

Lunardi, P., Santini, A., Crisà, M.: Il consolidamento della Rocca di Cefalù, Quarry and Construction (1994)

Marconi, P.: Girgenti. Ricerche ed esplorazioni. Notizie degli Scavi **II**, 93–148 (1926)

Marconi, P.: Cefalù (Palermo). Il cosiddetto «Tempio di Diana». Notizie degli Scavi **V**, 273–395 (1929)

Melis, M.G.: Osservazioni sul ruolo dell'acqua nei rituali della Sardegna preistorica, pp. 111–124. Rivista di Scienze Preistoriche, LVIII (2008)

Nott, G.F.: Avanzi di Cefalù. Monumenti Inediti dell'Instituto di Corrispondenza Archeologica, pp. 270–287, tavv. XXVIII-XXIX (1831)

Portale, E.C.: Le nymphai e l'acqua in Sicilia: contesti rituali e morfologia dei votivi. In: Calderone, A. (a cura di), Cultura e religione delle acque. *Atti del Convegno Interdisciplinare « Qui fresca l'acqua mormora.... » (S. Quasimodo, Sapph. fr. 2,5), Messina 29–30 marzo 2011*, Roma, pp. 169–191 (2012)

Remondino, F., Spera, M.G., Nocerino, E., Menna, F.: State of the art in high density image matching. Photogram. Rec. **29**(146), 144–166 (2014)

Saporetti, C.: Diana Facellina. Un mistero siciliano, Pungitopo Editrice, Marina di Patti (2008)

Tullio, A.: Saggio sulla topografia e sulle antichità di Cefalù. Kokalos **20**, 119–151, tavv. I-XVIIIV (1974)

Tusa, V.: *La Sicilia occidentale alla fine del V sec. ed agli inizi del IV sec. a.C.*, Palermo (1959)

Van Essen, M.C.C.: La date du sanctuarie mégalithique de Cefalù. Mélanges d'Archéologie et d'Histoire **59**, 45–48, tav. I-II (1957)

New Insights on the Akragas' Complex of Demeter and Persephone: The Role of the Moon

Robert Hannah, Giulio Magli and Andrea Orlando

Abstract We present here in details the results obtained—within a campaign held in August 2015, aimed at a complete re-evaluation of the temples of Akragas—on the Temple of Demeter and Persephone. This temple, built on the eastern slopes of the Athena Rock, belongs to the final phase of the Archaic period (480–470 BC) and is traditionally attributed to the Eleusinian divinities. The archaeoastronomical analysis hints at a connection with the Moon and opens up new perspectives in the problem of the attribution of the temple. In fact, the eastern front yields a declination very close to −27°, and therefore the building is one of the rare examples of Greek temples which is not oriented within the arc of the rising sun. Furthermore, the western front—which looks towards the Akragas acropolis—yields a declination very close to that of the Moon at the maximal northern standstill. A vast, artificial esplanade was built on this side and was very probably the place where nocturnal processions gathered after ascending to the temple.

1 Introduction

The Greek colonization of Sicily, started in the 7th century BC, led to the foundation of a series of towns such as Zancle, Naxos, Syracuse, Gela and Akragas (Agrigento). Together with the foundation and development of the colonies, the Greeks exported and developed sacred architecture, and so dozens of magnificent temples were constructed. The orientation of a Greek temple is defined as the direction of the main axis from the inside looking out, which is the direction in

R. Hannah
University of Waikato, Hamilton, New Zealand

G. Magli (✉)
Politecnico di Milano, Milan, Italy
e-mail: giulio.magli@polimi.it

A. Orlando
Catania Astrophysical Observatory (OACT/INAF), Laboratori Nazionali del Sud (LNS/INFN), Institute of Sicilian Archaeoastronomy (IAS), Catania, Italy

A. Orlando (ed.), *The Light, The Stones and The Sacred*,
Astrophysics and Space Science Proceedings 48,
DOI 10.1007/978-3-319-54487-8_6

which the statue of the god was in principle looking, as well as being the direction along which—eventually—the sun would illuminate the facade, which was the scenario for rites and celebrations taking place outside the temple at the altar.

The majority of Greek temples actually face the eastern horizon, mostly within the arc of the rising sun (e.g.: Boutsikas 2009 and references therein). In the case of the Greek temples of Sicily this pattern is particularly clear (Aveni and Romano 2000; Salt 2009) as 38 out of 41 measured temples are oriented within the arc of the rising sun. In the data samples of these authors, the temples out of the solar arc are: the Hellenistic sanctuary of Megara Hyblaea (7th century BC), the Asklepieion of

Fig. 1 Map of Akragas with temples and city walls (from Freeman 1891)

Eloro (4th century BC), the little temple A of Naxos (8th century BC), the Aphrodite temple of Heraclea Minoa (date unknown) and the Hekate temple of Selinous (mid 6th century BC).

Recently, the present authors have carried out an exhaustive campaign to measure all the temples of the UNESCO site of the Valley of the Temples in Akragas, Sicily (Hannah et al. 2016a, b). During this campaign we found out that, among the 10 temples of Akragas (Fig. 1), one monument had been almost completely forgotten from the archaeoastronomical point of view: the so-called temple of Demeter and Persephone. To our surprise, the temple turned out to be oriented outside the solar arc, and—as will be shown below—actually pointing to the Moon at the maximal northern standstill. In the present paper, a detailed discussion of this result and of its archaeological context is given. In particular, a long-standing archaeological question is the relationship of the temple with the so-called Archaic Fountain, a rather strange building, clearly connected with the cult of water, which stands just below the ridge of the temple.

Our results point to the existence of a deep connection between the two and perhaps to a nocturnal processional cult connected with the full moon which involved an ascension uphill from the fountain to the temple.

2 The Temple of Demeter and Persephone and the Archaic Fountain at Akragas

Akragas—today's Agrigento—was one of the most important Greek colonies in Sicily, founded in 582 BC by settlers from nearby Gela and from Rhodes. The site lies on a huge plateau, naturally protected from the north by the Athena Rock and the Girgenti Hills, and from the south by a long rib-hill, bounded on either side by the rivers Akragas and Hypsas, confluent to the south in a single water's course, at the mouth of which the port was constructed.

The Temple of Demeter and Persephone is located on the hills to the south-east of the town, immediately inside the city walls. It was later incorporated into the Norman church of San Biagio, which was built over it. However, the Doric temple is easily discernible (Fig. 2). It belongs to the final phase of the Archaic period (480–470 BC). The facade of the church points to the North-West, but it is likely that this facade was obtained by opening an entrance in the back wall of the cella of the temple, which was therefore originally facing South-East.

However, of the portico with two columns that the Greek temple should have had, only the (so-called honeycomb) foundation sectors can be seen (Fig. 3). The stone foundation platform of the whole building and a part of the original stone walls are preserved, accurately built in isodomic masonry of huge rectangular blocks. Inside the church, excavations have revealed a cistern belonging to the Greek phase, located close to the North-West corner and therefore inside the cella of the temple, a quite unusual feature (Bellavia et al. 2012). We retain some doubts that the original

Fig. 2 The apse of the Norman church of San Biagio above the remains of the Greek temple (authors' photo)

facade really was pointing to the South-East. Perhaps the excavators simply followed the paradigm that a Greek temple in Sicily is usually oriented to sunrise and overlooked the opposite possibility (the more because the temple is not oriented to sunrise anyway, see further below).

In any case there was no available space for an altar to the east, since the terrain slopes down abruptly on to the escarpment of the city walls. A large esplanade is instead present on the back (that is, in front of the church). This esplanade is contemporary with the temple and was obtained artificially through the construction of huge retaining walls on the south side and an accurate excavation and leveling of the rock on the north side. The area was accessed from the town through a large road partly excavated in the rock, which is still today perfectly visible.

The attribution of the temple was made possible by the discovery, during the excavations of 1925, of a votive deposit, formed of a large amount of objects. In particular, there were many fragments of two female busts of terracotta, one of which could be reconstructed in its integrity and was identified as Persephone. As a result of this discovery the temple is attributed to the Eleusinian divinities. This was also confirmed by the presence of two small circular altars (Fig. 4): one is solid, with a diameter of 2.53 m, and the other—with a diameter of 2.70 m—has a central well (*bothros*), which was found filled with ritual offerings, i.e. broken *kernoi*, or ritual vessels of Demeter.

These altars are located in the "corridor" formed between the rock cut to the north and the side of the temple.

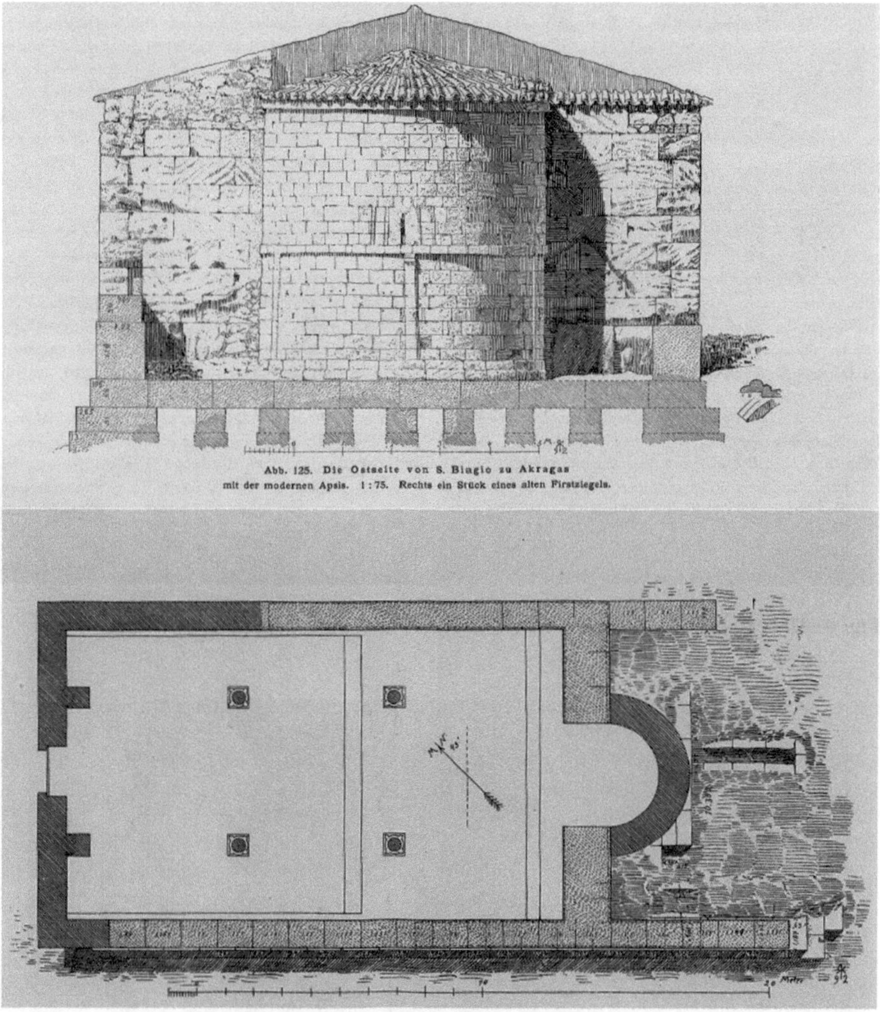

Fig. 3 Elevation (*above*) and plan (*below*) of the Norman church of San Biagio (from Koldewey and Puchstein 1899)

The temple could be accessed from outside the town by a stairway that crosses the town's walls through a postern and leads to a strange building located outside the walls. It is a protohellenic (7th century BC) sanctuary probably dedicated to chthonic deities, whose architectural elements are integrated with the natural features of the site, as is often found in the shrines of the gods of the earth (for example at Eleusis, in Lykosoura or in Enna). The sanctuary consists of a rectangular building up on the cliff (below the temple of Demeter and Persephone), on which there are two communicating hypogea, which were filled with votive offerings (Figs. 5 and 6). A third gallery was used as an aqueduct to supply water collected

Fig. 4 The two circular altars on the northside of San Biagio Church (authors' photo)

Fig. 5 The south-gallery of the rocky sanctuary (authors' photo)

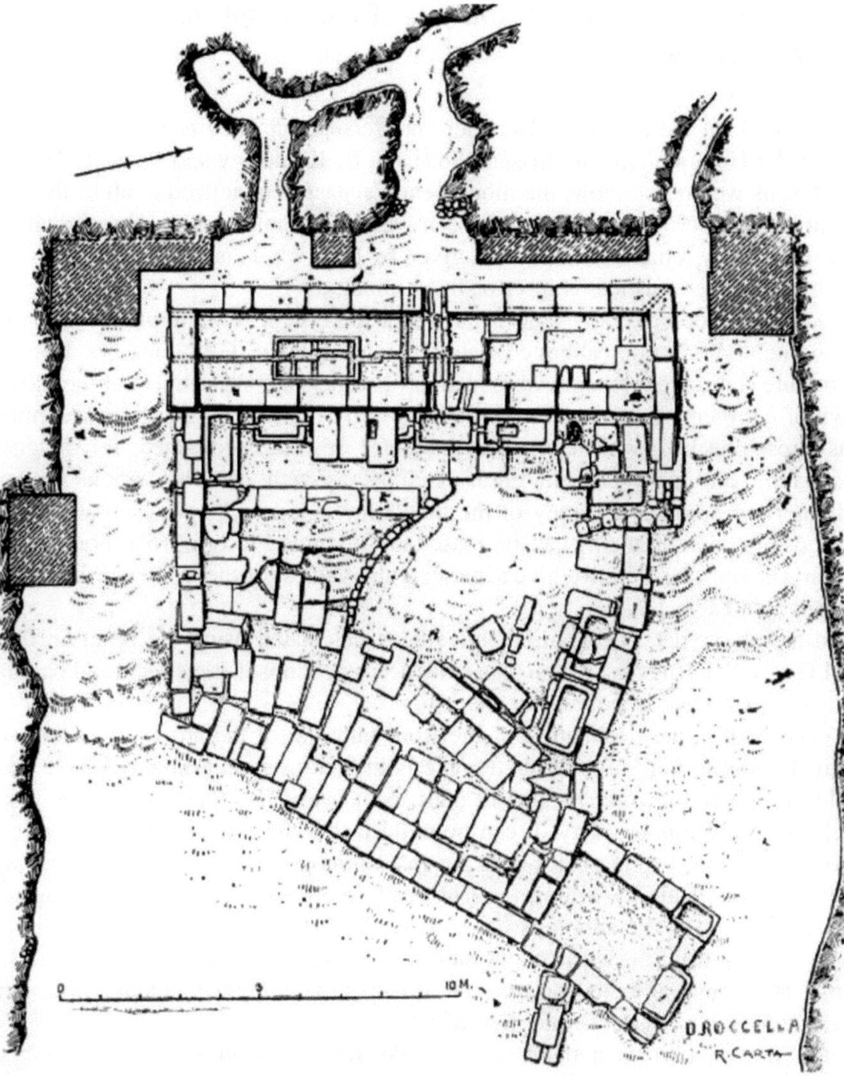

Fig. 6 The general plan of the archaic fountain (from Cultrera 1942)

from a nearby source in the basin of the building, which was therefore a sort of fountain-sanctuary (Zoppi 2004). Recently it has been proposed to associate the rock sanctuary with cults of the nymphs (Portale 2012) and new studies are underway to better understand its enigmatic architecture (Fino 2014).

3 Archaeoastronomical Analysis of the Temple of Demeter and Persephone

The azimuth of the temple of Demeter and Persephone was measured in the 19th century by Heinrich Nissen (Nissen 1869) and by Koldewey and Puchstein (1899); for reasons we do not know, the monument has passed unnoticed in all of the more recent works. Curiously, although Nissen was a very conscientious scholar and his data are usually reliable, the azimuth given by him, which is the same as the rising sun at the winter solstice, is in error by as much as 4°.

Our data, reported in Table 1, are measured with a high precision optical theodolite, with north calibrated using a long-distance GPS measure from the theodolite station to a recognizable feature (a corner of a skyscraper) of the modern town of Agrigento (at distances of about 2.5 km); consistency of all measurements was cross-checked with compass-clinometer readings corrected for magnetic declination and with Google Earth readings as well (due to various reasons, although the nominal accuracy of the theodolite is below 1′, we estimate that the error of our measurements can be reasonably assumed to be ±15′). For a reason which will soon be clear we give separately the values for the front and the back of the temple (Table 1).

In the above data, declinations have been calculated using the program DEC-CALC by Clive Ruggles, which takes into account atmospheric effects and parallax, which is extremely important for the Moon.

Very clearly, the declination of the front falls out of the arc of the rising sun. Actually, and to our surprise, the temple is the only one in Akragas whose declination is not in the solar range and adds to the very few Sicilian temples whose declination has this property. It can be noticed that the axis of the building could have been skewed some 4° towards the east on its platform, in order to align with winter solstice sunrise, without any practical problem. Even more, although the horizon in front of the temple is flat, a rock curtain located immediately to the left (east) looking from the entrance was left in situ (the huge excavation of the terrace ends just nearby), and this curtain even obscures the midwinter sun at rising. Thus, orientation was very deliberate. But why?

The first fact to be noticed is that a declination about −27° lies within the declinations of the sun at winter solstice and that of the moon at the major southern standstill, and fits the Venus minimal declination very precisely. We were therefore at first intrigued by the idea that the temple could be aligned to Venus. However,

Table 1 Data of azimuth, horizon, declination and lunar declination for the Temple of Demeter and Persephone, frontside and backside

	Azimuth (°)	Horizon (°)	Declination (°)	Lunar decl. (°)
Front	125.00	0	−26.53	−27.32
Back	305.00	2.38	+28.44	+29.22

although Venus can attain—in principle—its maximal and minimal declinations both as the morning and as the evening star, the *morning* star has never had a declination significantly greater than that of the sun at the solstices in the last four millennia or so. This fact is already well known from studies on Venus alignments in the Mayan world (Sprajc 1993), but in any case we verified it explicitly and independently in all of the 8-year Venus cycles that occurred in the 5th century BC. No conspicuous stars or asterisms correspond to this declination in this period either.

We then re-analyzed the orientation taking into account the possibility of a lunar alignment. Lunar declinations are affected by parallax by some ½° and therefore the front of the temple yields a lunar declination −27° 32′, not a significant value. At this point, we visited the temple again in search of an explanation. In our new visit we accurately measured the horizon to the North-West from the facade of the medieval church and therefore from the back side of the temple (Fig. 7). This horizon is very striking, since it is occupied by the hill where the acropolis of Akragas once sat. The tower of the medieval cathedral, which with all probabilities corresponds to the main temple on the acropolis, is clearly visible directly in front of the temple. Measuring the horizon we took into account an estimated average for the height of the modern buildings, and the results in declination are +28° 44′ uncorrected by parallax, with a lunar declination +29° 22′. This result is impressively close to the maximal lunar declination which in the 5th century BC (due to the slight variation of the obliquity of the Ecliptic) was around 28° 50′. Thus, we propose that the building was orientated to the setting of the Moon at the maximal northern standstill.

Fig. 7 The facade of the medieval San Biagio Church (authors' photo)

From the point of view of the cults practised at the temple, it may be noted that some authors have questioned the original dedication of the temple to Demeter and Persephone, attributing the scant archaeological evidence to a spurious origin (e.g. Bellavia et al. 2012). If this is true then, in our view, the peculiar location of the temple might instead point to an attribution to Artemis, a goddess associated both with boundaries and with the moon. Indeed Hesiod (ca. 700 BC) already connects Artemis closely with other astral deities, including Selene and Helios (Hesiod, *Theogony* 371–372, 404–412); the same occurs in the 5th century BC, in a fragment of Aeschylus (fr. 87) and is later well-established by the Hellenistic period, when, for instance, the 3rd–2nd century BC Stoic philosopher Diogenes of Babylon identified the moon with Artemis and the sun with Apollo (Cook 1940: 726 n.6).

Artemis' sanctuaries were predominantly associated with wild areas of Greece (Cole 2004: 180; Brulotte 1994) but many were also located at city entrances or the intersections of roads, e.g. those at Sikyon and Thebes (Pausanias 2.11.1, 9.17.1). Artemis was sometimes given the epithet *Enodia* ('in the road'), suggesting a link with roads as places between destinations (Johnston 1990 24 n. 10). The same epithet was given also to Selene, which suggests that this title of *Enodia* was closely associated with the Moon in some way. Artemis's epithets further suggest a relationship with boundaries. One was *Propylaia*, a title sometimes attached to the goddess at her entranceway temples, as at Eleusis (Clinton 1992: 116).

4 Discussion and Conclusions

As is well known, due to the interplay between positions at the horizon and lunar phases, precise lunar extreme azimuths are very difficult to individuate, but the full moon near the winter solstice in the years close to the standstill attains an azimuth which is always very close to the maximal one. The midwinter full moon rises at sunset and sets close to sunrise, remaining in the sky almost the whole night. This leads us to suppose the existence of a rite connected with this recurrence. Nocturnal rites are known from other parts of the Greek world from an early date. Alkman (mid-7th century BC) refers to a ritual that took place just before dawn at the time of the heliacal rising of the Pleiades (Alkman, Partheneion 39–43, 60–63; Boutsikas and Ruggles 2011, 60–65).

Aristophanes famously notes the sighting of the morning-star ('the light-bearing star of our nocturnal rite', Aristophanes, *Frogs* 342–343) at the beginning of the fifth day of the Eleusinian Mysteries (Boutsikas and Hannah 2012). Also in Athens the Kallynteria, the Plynteria and the Arrhephoria took place on the Acropolis in Athens and were associated with observations of the stars just before dawn. On Keos the islanders offered sacrifices to Sirius at its heliacal rising (Apollonius, *Argonautica* 2.516–27). In addition, the Eleusinia in Athens involved a nighttime procession between Athens and Eleusis. So it is conceivable that processional rites were carried out at Akragas, which involved both the fountain sanctuary and the temple uphill. We can imagine a nocturnal procession coming up from the sanctuary and reaching the temple, in front of which, however, there is not—and there never has been—enough space to house worshipers. People attending the procession, after the ascent, crossed the corridor between the north side of the temple and the hill—perhaps throwing votive offerings in the bothros—and gathered in the vast esplanade located on the back of the temple.

The setting of the full moon over the acropolis of Akragas at dawn near midwinter would have certainly been a unforgettable spectacle for all of them.

Acknowledgements The present work has been developed within the convention for scientific cooperation existing between the "Ente Parco Archeologico della Valle dei Templi" and the FDS Laboratory for the Communication of Science at the Politecnico of Milan. The authors wish to thank Giuseppe Parrello, Dirretore di Ente Parco, and Dr. Giuseppe Presti, Responsabile di Ente Parco, for their cooperation and support.

References

Aveni, A.F., Romano, G.: Temple orientations in Magna Graecia and Sicily. J. History Astron. **31**, 51–57 (2000)

Bellavia, M., et al.: Nuovi dati sulla Chiesa di San Biagio ad Agrigento. Archeologia Postmedievale **16**, 125–132 (2012)

Boutsikas, E.: Placing Greek temples: an archaeoastronomical study of the orientation of ancient Greek religious structures, in archaeoastronomy. J. Astron. Culture **21**, 4–19 (2009)

Boutsikas, E., Ruggles, C.: Temples, stars, and ritual landscapes: the potential for archaeoastronomy in ancient Greece. Am. J. Archaeol. **115**(1), 55–68 (2011)

Boutsikas, E., Hannah, R.: Aitia, astronomy and the timing of the Arrhēphoria. Annu. Br. School Athens **107**, 233–245 (2012)

Brulotte, E.L.: The placement of votive offerings and dedications in the Peloponnesian sanctuaries of Artemis. Ph.D. Dissertation, University of Minnesota (1994)

Clinton, K.: Myth and cult: the iconography of the Eleusinian mysteries. The Martin P. Nilsson Lectures on Greek Religion delivered 19–21 November 1990 at the Swedish Institute at Athens, Paul Aströmsförlag, Stockholm (1992)

Cole, S.G.: Landscapes, Gender and Ritual Space: The Ancient Greek Experience. University of California Press, Berkeley (2004)

Cook, A.B.: Zeus: A Study in Ancient Religion, vol. 3. Cambridge University Press, Cambridge (1940)

Cultrera, G.: Il santuario rupestre presso San Biagio in Agrigento, in Atti AccPalermo, s. IV, III, pp. 609–627 (1942)

Fino, A.: La fontana arcaica di San Biagio ad Agrigento. In: Livadiotti, M., Parello, M.C. (a cura di), Il restauro dei monumenti antichi. Problemi strutturali: esperienze e prospettive, Atti delle Giornate di Studio, Agrigento, 23–24 novembre 2012, Thiasos, 3.2, Convegni, pp. 67–91 (2014)

Freeman, E.: The History of Sicily, p. 224. Clarendon Press, Oxford (1891)

Hannah, R., Magli, G., Orlando, A.: The role of urban topography in the orientation of Greek temples: the cases of Akragas and Selinunte. Mediterr. Archaeol. Archaeom. **16**(4), 213–217 (2016a)

Hannah, R., Magli, G., Orlando, A.: Astronomy, topography and landscape at Akragas' Valley of the Temples. J. Cult. Herit. (2016b) (in press)

Johnston, S.I.: Hekate Soteira: A Study of Hekate's Roles in the Chaldean Oracles and Related Literature. Scholars Press, Atlanta (1990)

Koldewey, R., Puchstein, O.: Die griechischen Tempel in Unteritalien und Sicilien, Berlin (1899)
Nissen, H.: DasTemplum. Antiquarische Untersuchungen, Berlin (1869)
Portale, E.C.: Le nymphai e l'acqua in Sicilia: contesti rituali e morfologia dei votivi. In: Calderone, A. (a cura di), Cultura e religione delle acque. Atti del Convegno Interdisciplinare «Qui fresca l'acqua mormora….» (S. Quasimodo, Sapph. fr. 2,5), Messina 29–30 marzo 2011, Roma, pp. 169–191 (2012)
Salt, A.M.: The astronomical orientation of ancient Greek temples. PLoS ONE **4**(11) (2009)
Sprajc, I.: Venus orientations in ancient Mesoamerican architecture. In: Ruggles, C.L.N. (ed.) Archaeoastronomy in the 1990s. Group D Publications, Loughborough (1993)
Zoppi, C.: Le fasi costruttive del cosiddetto santuario rupestre di San Biagio ad Agrigento: alcune osservazioni, Sicilia antiqua. Int. J. Archaeol. **I**, 41–79 (2004)

A Study on the Orientation of Greek Theatres

Marzia Monaco, Flavio Carnevale and Marcello Ranieri

Abstract This study on the orientation of Greek theatres has been performed measuring on satellite images the azimuth of a sample of 84 theatres divided in three geographical groups (*Magna Grecia* and Sicily, Mainland Greece and Greek theatres elsewhere). Comparison with the values that can be found in literature (achieved on place with the compass, often without taking into account the magnetic deviation, or derived from archaeological plans) shows that measures performed on satellite images prove to be more reliable. The main purpose was to determine whether Greek theatres present any preferential orientation and, when possible, to investigate the reasons. From our study a preference for a Southern orientation clearly appears. Using SunCalc we have considered the impact of the solar exposure on the South-facing theatre of *Catina* in Sicily. The most probable explanation is that a southern orientation allows the maximum natural lighting and warming for the *cavea* over the whole day during all the seasons of the year.

1 Introduction

Although the number of catalogues of Ancient Greek and Roman theatres is relevant, the majority only reports generic indications on the orientations such as South-facing, North-facing, etc. Orientations can be found in literature in single cases, but studies on a consistent sample restrict substantially to those by Von Gerkan and Müller-Wiener (1961) and by Ashby (1999).

Von Gerkan and Müller-Wiener presented a case study on the theatre of *Epidaurus* in which a diagram shows the azimuth of 28 theatres with numerical values reported there on (Fig. 1).

M. Monaco (✉) · F. Carnevale
Rome «La Sapienza» University, Rome, Italy
e-mail: marzia.monaco@uniroma1.it

M. Ranieri
INAF-IAPS, Rome, Italy

A. Orlando (ed.), *The Light, The Stones and The Sacred*,
Astrophysics and Space Science Proceedings 48,
DOI 10.1007/978-3-319-54487-8_7

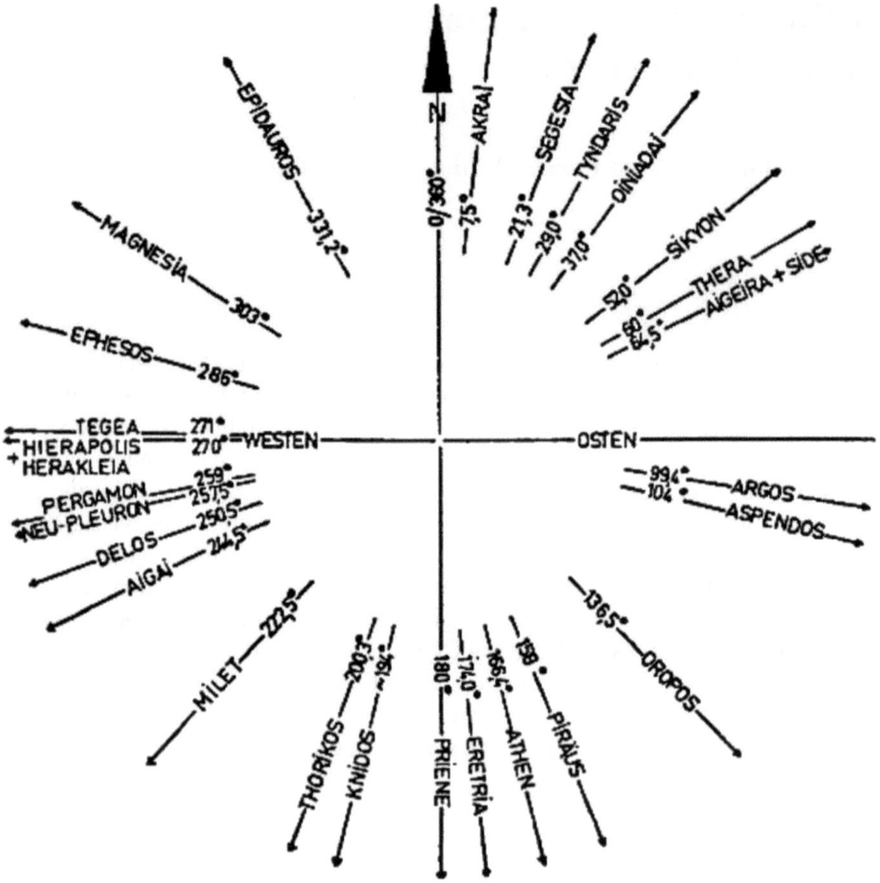

Fig. 1 Diagram of the theatres orientation (from Von Gerkan and Müller-Wiener 1961)

The work by Ashby, in which an entire chapter is on orientations, is the one that includes the higher number of Greek theatres. Ashby divides his sample of 123 theatres in four groups (*Magna Grecia*, Mainland, Islands and Anatolia). Bearings are not reported numerically on his diagrams, but percentages are given in a table (Fig. 2).

These studies are however quite aged. The work by Von Gerkan and Müller--Wiener is 50 years old. The one by Ashby was published 17 years ago, and, as the author declares, the study began 16 years before for a total of more than 30 years. Ashby also explains that:

> ...bearings are not of equal reliability; many are taken from such and undoubtedly correct sources as excavation ground plans, but a few are from dubious free hand sketches found in various guidebooks. a large number come from personal observations made at the sites with declination disregarded...

	North (315–44°)		East (45–134°)		South (135–224°)		West (225–314°)		Total
Mainland Greece	18%	(9)	24%	(12)	43%	(21)	14%	(7)	49
Anatolia	15%	(7)	21%	(10)	48%	(23)	17%	(8)	48
Greek Islands	8%	(1)	42%	(5)	25%	(3)	25%	(3)	12
Magna Graecia	14%	(2)	29%	(4)	50%	(7)	7%	(1)	14
Total	15%	(19)	25%	(31)	44%	(54)	15%	(19)	123

Fig. 2 Percentages of the theatres orientations (from Ashby 1999)

It should be added that many of his bearings were derived from *Teatri classici in Asia Minore* by De Bernardi Ferrero (1966-74).

For a complete inventory of all the known Greek theatres we referred to the list by Pedersoli (2015) who cross-checks all theatres appearing in previous studies and updates with those most recently excavated. In his list theatres classified as Greek or Greek-Roman (i.e. Roman theatres edified over earlier Greek ones) amount to 232. This number includes 28 no more visible theatres (because destroyed, not excavated or covered after the excavation), 70 identified only on literary, topographic or archaeological bases, and 134 still visible. Excluding 11 rectilinear structures for which the identification with a theatre is dubious, the sample reduces to 123 visible circular theatres.

2 Methodological Approach

Orientations can be determined with on place measurements or derived from archaeological plans or from satellite images. To retrieve at the best the orientations from archaeological plans an adequate methodology should be followed, as in Ranieri (2014). The most reliable plan should be chosen among all those available from archaeological sources. Misinterpretations may occur on the significance of the North arrow which may indicate True North or Magnetic North. In some cases it is specified, occasionally with the indication of the related date of the measurement, but mostly without. Sometimes, especially on the oldest plans, both True and Magnetic North arrows can be found, either with or without date and value of the magnetic deviation. The knowledge of the date is essential to recover the magnetic deviation from the Historical Geomagnetic data of NOAA (National Oceanic and Atmospheric Administration).

This procedure would result greatly time consuming for this work. Also time consuming would have been to measure the orientations on place. Therefore we decide to use satellite images. Nowadays they are available for free on the web (Google Earth and Bing Maps); the resolution is appropriate to our purposes and no knowledge of the magnetic deviation is necessary. Problems may only occur if the

Fig. 3 The 84 still visible circular theatres, the original outline of which is clearly recognizable on satellite images (elaborated on Google Earth)

images of the theatres result partially covered by clouds (as for instance in the case of *Cheronea*) or because of their poor quality.

Of the 123 still visible circular theatres, around 90 could be localized on Google Earth. Since the recognisability of the outline is essential to determine the orientation, six theatres had to be discarded while 84 were considered suited for a graphical study (Fig. 3). The sample of 84 theatres was divided in three geographical groups: theatres of *Magna Grecia* and Sicily (16), theatres of Mainland Greece (29), and Greek theatres elsewhere (39), that is Albania, Bulgaria, Cyprus, Iraq, Libya and Turkey.

Satellite images were acquired with screen captures from a high-resolution screen and imported in.jpg format in Geogebra application.[1]

Best-fit procedures were used to recognize the form and to identify the axis of the theatres. For any theatre, the circumference that best-fits the *cavea* allows to mark the most probable centre. The identification of the best fitting line matching the *scaena* allows to determine the orthogonal line passing through the centre, that is the axis of the theatre (Fig. 4a). The bearing was determined measuring the angle from North to the axis in clockwise direction (from North to East), being the direction intended from the *cavea* towards the *scaena* (Fig. 4b). The error on the bearings has been estimated to be ±1.5°.

[1]GeoGebra is an interactive geometry, algebra, statistics and calculus software.

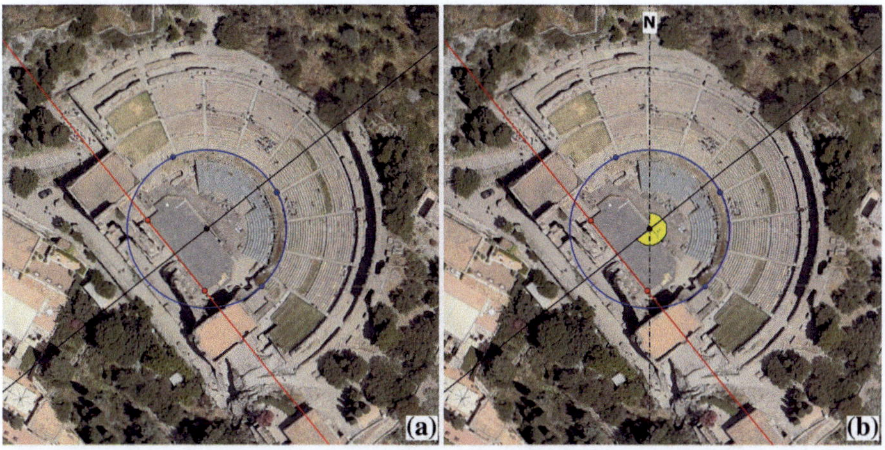

Fig. 4 *Tauromenium* determination of the axis **a** and measure of the azimuth **b**

Table 1 Azimuth values measured for the 16 theatres of *Magna Grecia* and Sicily. Legenda: AZ = Azimuth measured on satellite images, AZ_{AS} = Azimuth measured on the diagram by Ashby, Δ = difference, G = Greek theatres, G/R = Greek/Roman theatres

ID	SITE	TYPE	DATE	AZ (°)	AZ_{AS} (°)	Δ (°)
1	Catina	G/R	end V cent. B.C.	**174.6**	191.3	−16.7
2	Heraclea Minoa	G	IV–III cent. B.C.	**177.0**	175.2	1.8
3	Romechium	G/R	II–I cent. B.C.	**146.6**		
4	Locri Epizephyrii	G	IV cent. B.C.	**126.6**	124.3	2.3
5	Metapontum	G	IV–III cent. B.C.	**166.7**	180.0	−13.3
6	Iaitas	G	IV cent. B.C.–200 B.C.	**183.0**		
7	Acrae	G	II cent. B.C.	**8.4**	0.0	8.4
8	Pompeii	G/R	II–I cent. B.C	**144.9**		
9	Hippana	G		**329.7**		
10	Segesta	G	half III cent. B.C.	**11.2**	210.4	−199.3
11	Morgantina	G	IV–III cent. B.C.	**69.7**	73.8	−4.1
12	Syracusae	G	IV–III cent. B.C.	**181.8**	186.4	−4.6
13	Solus	G	IV cent. B.C.	**113.4**	110.4	3.0
14	Tauromenium	G	II cent. B.C.	**230.9**	232.8	−2.0
15	Tyndaris	G	300 B.C.	**37.4**	28.3	9.1
16	Velia	G	IV cent. B.C.	**126.3**	120.4	5.9

2.1 *Greek Theatres of* Magna Grecia *and Sicily*

Our measurements of the azimuth for the 16 theatres of *Magna Grecia* and Sicily are shown in Table 1. A comparison with those by Ashby is possible for 12 theatres. The distributions of the values might appear similar, but at deeper insight

Table 2 Azimuth values measured for the 29 theatres of Mainland Greece. Legenda: AZ = Azimuth measured on satellite images, AZ_{AS} = Azimuth measured on the diagram by Ashby, Δ = difference, G = Greek theatres, G/R = Greek/Roman theatres

ID	SITE	TYPE	DATE	AZ (°)	AZ_{AS} (°)	Δ (°)
17	Argus	G	300–275 B.C.	**106.8**	99.3	7.6
18	Athenae	G	325 B.C.–II cent. B.C.	**161.3**	172.2	−10.9
19	Corinthus	G/R	IV cent. B.C.–I cent. B.C.	**356.8**	357.7	−0.9
20	Delphi	G	End III cent. B.C.–160 B.C.	**138.4**	158.7	−20.3
21	Delus	G	End IV cent. B.C.–246 B.C	**254.5**	250.7	3.8
22	Dion	G	200 B.C.	**59.3**	85.8	−26.5
23	Dodona	G	297–272 B.C.–168 B.C.	**159.4**	153.9	5.4
24	Aegira	G	280–250 B.C.	**59.9**	56.7	3.2
25	Elis	G	End IV cent. B.C.	**318.4**	309.4	8.9
26	Epidaurus	G	Beginning III cent. B.C.	**330.3**	330.6	−0.3
27	Eretria	G	IV cent. B.C.	**179.7**	190.4	−10.7
28	Poseidon	G	400 B.C.	**9.1**	6.9	2.2
29	Oropus	G	IV cent. B.C.–200–150 B.C.	**135.8**	136.2	−0.4
30	Cassope	G	Beginning III cent. B.C.	**152.1**	155.8	−3.7
31	Nea Pleuron	G	III cent. B.C.	**273.0**	257.8	15.2
32	Oeniade	G	III cent. B.C.	**210.8**	224.4	−13.6
33	Philippi	G/R	post 356 B.C.	**168.2**	167.0	1.2
34	Larisa	G	End III cent. B.C.	**186.6**	176.1	10.5
35	Hephaestia	G	II cent. B.C.	**160.6**	195.0	−34.4
36	Maronea	G		**235.3**	225.2	10.1
37	Messene	G/R	II cent. B.C.	**181.4**	110.3	71.1
38	Megalopolis	G	370 B.C.	**11.0**	11.9	−0.9
39	Orchomenus	G	End IV cent. B.C.	**91.7**	88.2	3.4
40	Sicyon	G	III cent. B.C.	**52.7**	54.7	−2.0
41	Sparta	G	I cent. B.C.	**192.5**	195.7	−3.2
42	Thasus	G	IV–III cent. B.C.	**299.5**	304.8	−5.3
43	Thoricus	G	500 B.C.–IV cent. B.C.	**201.8**	198.4	3.4
44	Mantinea	G	IV cent. B.C.	**88.1**	90.7	−2.6
45	Demetrias Pag.	G	Beginning III cent. B.C.	**117.6**	119.9	−2.4

we notice some relevant differences. The theatre of *Segestae* is reported by Ashby in the opposite direction (possibly a misreading of the compass). An important amount of error (more than 10°) is present on 25% of the cases. In the case of *Catina* the difference is more than 15°: Ashby reports that the bearing was taken on the remains of the foundation of the original Greek theatre and possibly the large error depends on his difficulty to ascertain the complete outline of the theatre.

Table 3 Azimuth values measured for the 39 Greek theatres elsewhere. Legenda: AZ = Azimuth measured on satellite images, AZ_{AS} = Azimuth measured on the diagram by Ashby, Δ = difference, G = Greek theatres, G/R = Greek/Roman theatres

ID	SITE	TYPE	DATE	AZ (°)	AZ_{AS} (°)	Δ (°)
46	Buthrotum	G	III cent. B.C.	**161.3**	158.1	3.2
47	Phoinike	G	III–II cent. B.C.	**212.3**		
48	Byllis	G	III cent. B.C.	**285.6**	278.3	7.3
49	Nikaia	G	III cent. B.C.	**329.4**		
50	Apollonia [Pojani]	G	III cent. B.C.	**224.3**	280.2	−55.9
51	Philippopolis	G	IV cent. B.C. (?)	**185.0**		
52	Curium	G	II sec. B.C.	**215.8**		
53	Paphus	G	End IV cent. B.C.	**89.1**	189.9	−100.8
54	Babylon	G	III cent. B.C.	**172.0**		
55	Apollonia [Marsa Susa]	G	300–280 B.C.	**334.1**		
56	Cyrenae	G	IV cent. B.C.	**23.5**		
57	Magnesia ad M.	G	II-I cent. B.C.	**0.2**	305.8	54.5
58	Arycanda	G	II cent. B.C.	**168.2**	174.6	−6.4
59	Miletus	G	300 B.C.	**225.4**	223.8	1.5
60	Pergamum	G	III–II cent. B.C.	**254.2**	262.6	−8.5
61	Halicarnassus	G	II cent. B.C.	**173.0**		
62	Caunus	G	II–I cent. B.C.	**229.6**	238.9	−9.3
63	Stratonicea	G	II cent. B.C.	**359.8**		
64	Aphrodisia	G	II cent. B.C.	**174.6**	110.5	64.2
65	Cibyra	G	100 B.C.	**109.0**		
66	Laodicea ad Lyc.	G	II cent. B.C.	**37.9**		
67	Laodicea ad Lyc.	G	II cent. B.C.	**291.2**	284.1	7.1
68	Priene	G	300 B.C.	**181.1**	180.5	0.6
69	Termessus	G	II cent. B.C.	**104.4**	114.3	−9.9
70	Troia	G	300 B.C.	**181.3**		
71	Erythrae	G	IV cent. B.C.	**349.4**		
72	Alinda	G	II cent. B.C.	**138.0**	143.0	−5.0
73	Patara	G	I cent. A.D.	**29.3**	53.8	−24.5
74	Xanthus	G	III–II cent. B.C.	**8.8**	20.2	−11.4
75	Letoon	G	100 B.C.	**324.0**	330.2	−6.2
76	Iasus	G	II cent. B.C.	**39.6**	37.7	1.9
77	Pinara	G	II cent. B.C.	**269.0**	280.1	−11.1
78	Hierapolis	G	II cent. B.C.	**260.7**		
79	Rhodiapolis	G	II cent. B.C.	**175.1**	179.2	−4.1
80	Ephesus	G	II cent. B.C.	**283.2**	71.0	212.2
81	Nysa	G/R	ante 161 A.D.	**184.3**	176.1	8.3
82	Cnidus	G	II–I cent. B.C.	**195.8**		
83	Tlos	G	End I cent. B.C.	**303.5**	200.2	103.3
84	Cyaneae	G	III–II cent. B.C.	**184.2**		

2.2 Theatres of Mainland Greece

Our measurements of the azimuth of the 29 theatres of Mainland Greece are shown in Table 2. Comparing the values with those deduced from the diagram by Ashby, only 5 bearings are substantially coincident (differences less than 1°). For other 14 the differences are between 2° and 10°, but for the remaining 10 theatres the differences are large, ranging between 10° and 71°.

2.3 Greek Theatres Elsewhere

The values measured for the 39 Greek theatres elsewhere are shown in Table 3. Comparison with bearings by Ashby could be made for 24 theatres. Only one bearing (*Priene*) is substantially coincident, while large differences were found for the others. The bearing of *Ephesus*' theatre, reported by Ashby in his diagram, is about 71° in front of our measurement of 283° (a possible misreading of the

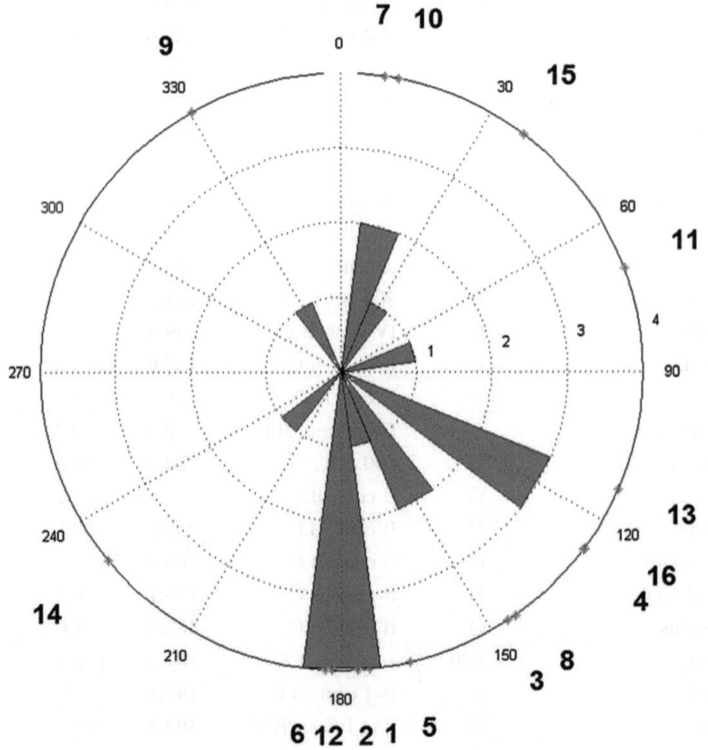

Fig. 5 Distribution of the 16 *Magna Grecia* and Sicily theatres

compass). The bearing of *Tlos* by Ashby is about 200° with a difference of 103° from our value. A similar difference occurs for *Paphus* (Ashby 190° in front of our 89°). A relevant amount of error (between 10° and 64°) is present on 6 theatres, while for the other 13 (54%) the differences are between 2° and 10°.

3 Discussion

From the examination of our measures it results a non random distribution of the orientations of Greek theatres. This reinforces the opinion of Ashby who judges untenable the often put forward "random hillside" approach according to which "the builders sought only a suitably graded slope in an accessible location" and that may come in mind looking at the aged diagram by Von Gerkan and Müller-Wiener (Fig. 1).

Ashby divides the directions cardinally in four sectors (Fig. 2). Since our evaluations of the bearings are more precise and reliable (±1.5°) we have constructed diagrams grouping the data in smaller, 15° wide, bins.

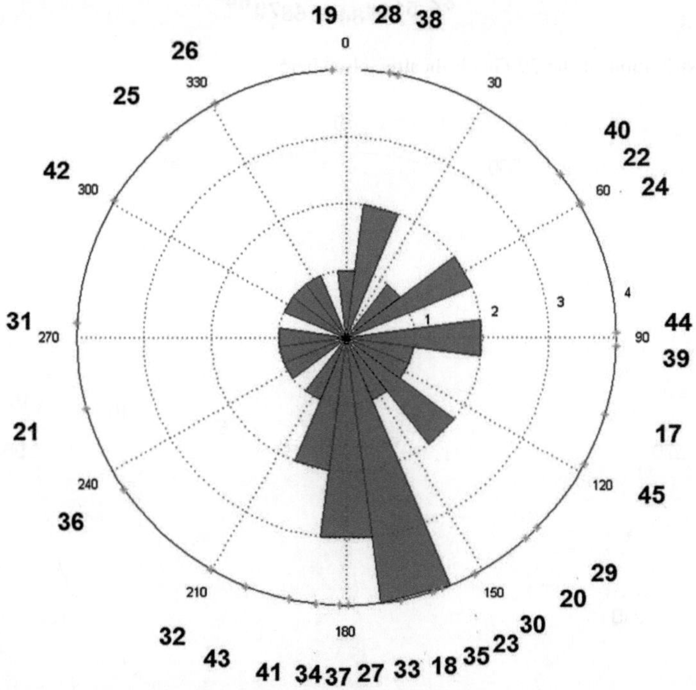

Fig. 6 Distribution of the 29 Mainland Greece theatres

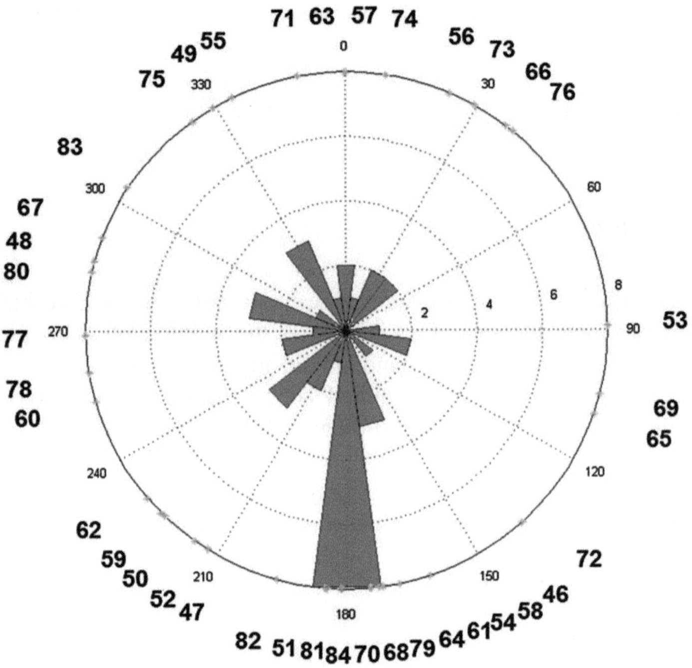

Fig. 7 Distribution of the 39 Greek theatres elsewhere

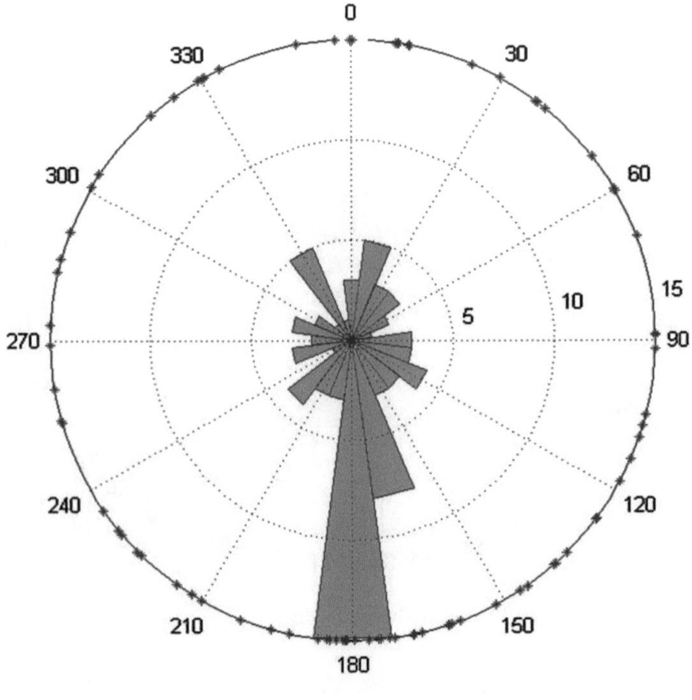

Fig. 8 Distribution of the 84 Greek theatres

For *Magna Grecia* and Sicily (Fig. 5) the richest two bins (three or more occurrences) represent about 44% of the total. The most populated is centred on due South (25%); the second most populated (18.8%) is centred on solstice (120°).

For Mainland Greece (Fig. 6), again a preference for a south orientation appears. The richest two bins represent about 24% of the total: the most populated is centred on 15° East of South (not far from South) and the second is on South (10.3%).

For Greek theatres elsewhere (Fig. 7) the richest bin represents 20.5% of the total. As for *Magna Grecia* and Sicily it is centred on due South. Also appears populated (7.7%) the bin centred on 15° East of South.

In all the three diagrams, with a statistically significant percentage, it emerges that a preference for South orientation do exists. As can be seen in Fig. 8, considering the whole sample of 84 theatres the most populated bins restrict essentially to two: the richest centred on due South (17.9%) and the other centred on 15° East of South (9.5%), for a total of 27.4%.

In the table by Ashby (Fig. 2) the percentage of theatres in the 90° wide South sector (135–225°) is 44%. Considering the same angular range, this percentage is confirmed by our data (37/84), but a consistent majority (23/37) is much more strictly South oriented (157.5–187.5°).

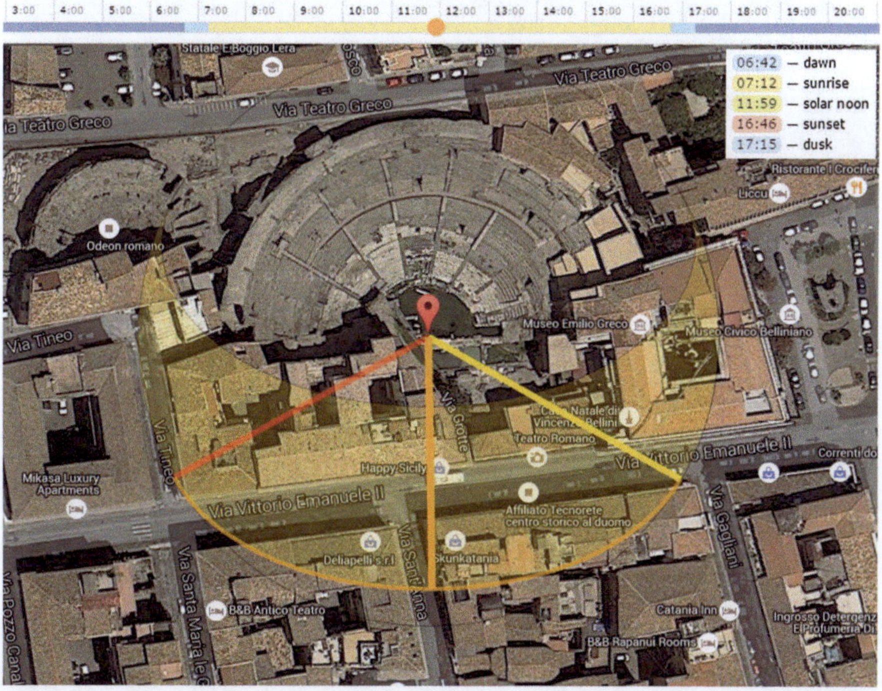

Fig. 9 *Catina* positions of the sun at Winter Solstice

4 The Southern Orientation: A Focus on the Theatre of *Catina*

According to Ashby (1999), the orientation of the theatres is related to the seasonal position of the sun. The author reports that:

> Theatres were sited to make the best use of the sun at particular times of the year… The southern exposure invite the sun to warm both spectators and benches; warmth was a primary concern in winter… but may be not desiderable at midsummer… Theatres were most likely positioned to provide suitable heat and lighting for the time of year when most important performances were presented.

Computer software can help further researches on the natural lighting of the monuments. Simulations of the solar motion may help in revealing the reasons that might have influenced the planning of the theatre. We used SunCalc, a web application working on 2D satellite images (Google Maps and Google Earth) by which the movement of the sun can be simulated. Specifically, the most relevant positions of the sun (sunrise, solar noon and sunset), the trajectory and its variation can be followed over the whole year.

As an example, we have considered the South-facing theatre of *Catina* for which is possible to highlight that:

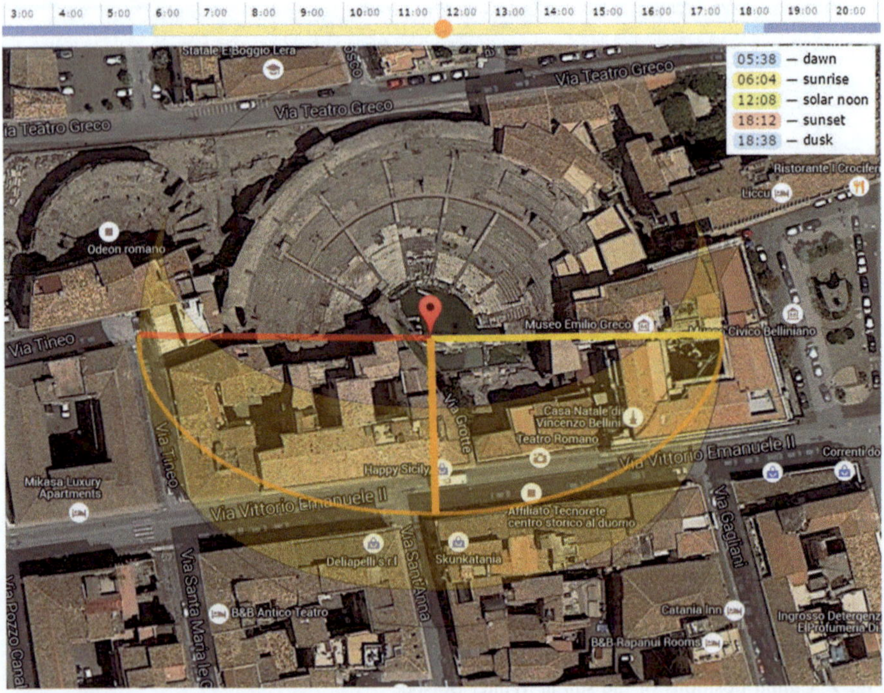

Fig. 10 *Catina* positions of the sun at the Equinoxes

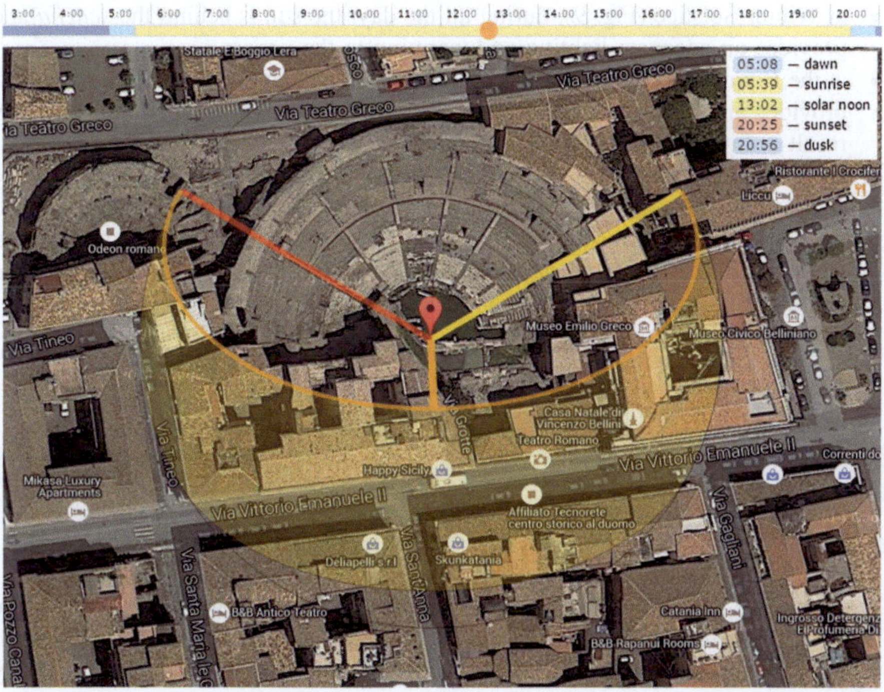

Fig. 11 *Catina* positions of the sun at Summer Solstice

Fig. 12 *Siracusae* Summer Solstice sunset (from Google Earth)

- at Winter Solstice (Fig. 9), during all the day, the sun is behind the *scaena* and provides an adequate lighting and warming to the *cavea*;
- at the Equinoxes (Fig. 10), during all the day, the sun is behind the *scaena* and provides an adequate lighting and warming to the *cavea* but, perhaps, this may result unfavourable in the hottest hours;
- at Summer Solstice (Fig. 11), for the large part of the day the sun is behind the *scaena* which is only illuminated from sunrise to 8:00 and from 17:00 to sunset; the opposite occurs for the *cavea*: the sun is behind only at sunrise and sunset, while for the rest of the day it provides an adequate lighting and warming (certainly unfavourable in the hottest hours).

Clearly this type of orientation allows the maximum natural lighting and warming for the *cavea* over the whole day, while the *scaena* would remain illuminated only if no high structures existed behind the players.

5 Conclusions

The study has shown that for large overground structures, clearly visible on Google Earth or Bing, the determination of the azimuth results more reliable if performed on satellite images. The measure of the orientation should be made on plans only after an accurate inspection of it and of all the related archaeological and historical sources. If it is not possible to determine with certainty the significance of the North arrow, the date of the drafting or the correlated magnetic deviation, then the measures of the azimuth might be not reliable.

Since theatres are public architectures mostly with functional purposes, the orientation assumes a different significance with respect to religious architectures. For religious architectures the orientation can have been intended to confer to the building a special symbolic significance related to astronomical (temples, sanctuaries, churches, etc.) or topographical/geographical reasons (as for mosques oriented toward Mecca). Vice versa, the orientation of the theatres appears more strictly linked to the advantages produced by the solar lighting on the spectators during those particular times of the year when the performances took place. In these cases quantitative and statistical analyses may be not sufficient to provide by themselves exhaustive explanations of the orientations. Necessarily they have to be combined with geographical, geo-morphological and topographical investigations and with critical studies of single theatres done by archaeologists and by historians of Ancient Greek theatre.

For future investigations Google Earth 3D models (when available) can be a precious tool. The dynamic views and the time control function allow to verify the effect of the direct lighting on the theatre during the day and the seasons (Fig. 12). Also 3D models obtainable with laser scanner and photogrammetric technologies, imported in a proper software (as Stellarium) or simply in CAD, can be useful for a simulation of the lighting of the architectures.

References

Ashby, C.: The orientation of Greek Theatres. In: Ashby, C. (ed.), Classical Greek Theatre, pp. 97–117. New Views of an Old Subject, University of Iowa Press, Iowa City (1999)

De Bernardi Ferrero, D.: Teatri classici in Asia Minore, Roma: L'Erma di Bretscheneider (1966–74)

Pedersoli, A.: Teatri Greci e Romani censiti (2015). http://www.engramma.it/eOS/index.php?id_articolo=385

Ranieri, M.: Digging the archives: the orientation of Greek Temples and their diagonals. MAA **14** (3), 15–27 (2014)

Von Gerkan, A., Müller-Wiener, W.: Das Theater von Epidauros. Kohlhammer, Stuttgart (1961)

Argimusco: Cartography, Archaeology and Astronomy

Andrea Orlando

Abstract Argimusco is a district in the neighborhood of the medieval city of Montalbano Elicona, in northern Sicily. It is a plateau, a suggestive naturalistic area that is near to many interesting archaeological sites, like Rocca San Marco and Sperlinga, whose origins date back respectively to the Paleolithic and Mesolithic Age (Brea in La Sicilia prima dei Greci, Il Saggiatore, Milano, 1958; Cavalier Il riparo della Sperlinga di S. Basilio (Novara di Sicilia), in Bullettino di Paletnologia Italiana, Roma, pp. 7–76, 1971). The Argimusco's plateau has natural rock formations featuring anthropomorphic and zoomorphic forms. It is assumed that the entire area had been used since the prehistoric times as a place of worship and ritual use, and, according to some sicilian scholars, it could be a sort of natural sanctuary (Pantano in Megaliti di Sicilia, Edizioni Fotocolor, Patti, 1994; Todaro in Allaricerca di Abaceno, Armando Siciliano Editore, Messina, 1992). Unfortunately, official archaeological excavations were never made on the Argimusco's plateau, but, from several surveys, lithic and pottery fragments have emerged ranging from the prehistoric age to the late medieval period. Argimusco appears to be a good site for the observation of the sky and of celestial phenomena (Orlando Archaeoastronomy in Sicily: Megaliths and Rocky Sites, in The Materiality of the Sky, Sophia Centre Press, 2016). In this paper I present the results of a 6 years of study of this area, and I propose a possible calendar based on the eastern horizon profile characteristics with Rocca Novara that acts as an equinox indicator.

A. Orlando (✉)
Catania Astrophysical Observatory (INAF), Catania, Italy
e-mail: aorlando@oact.inaf.it

A. Orlando
National South Laboratories (INFN), Catania, Italy

A. Orlando
Institute of Sicilian Archaeoastronomy, Novara di Sicilia, Italy

A. Orlando (ed.), *The Light, The Stones and The Sacred*,
Astrophysics and Space Science Proceedings 48,
DOI 10.1007/978-3-319-54487-8_8

1 Geological Framework

In NE Sicily the Peloritani Mounts, together with the Calabrian Costal Chain and Sila, Serre and Aspromonte Massifs, forms the orogenic domain known as the Calabria-Peloritani Arc (CPA). The CPA is an arcuate structure connecting the NW-SE trending Apennines to the E-W trending Sicilian and African Maghrebids, in the Africa-Adria-verging Mediterranean Alpine Chain.

The CPA is formed by a stack of continental and oceanic crust units. It was structured in the Tertiary during the last collisional phases between Europa and Africa Plates. In the Argimusco area the most important outcrops are the siliciclastic turbidites of the Capo d'Orlando Flysch (COD) and secondary the Floresta Calcarenites (CFL) and the Antisicilide Complex (ASI).[1] The COD is characterized by arenaceous-conglomeratic facies evolving upward and laterally to arenaceous-pelitic facies (ISPRA 2011).

The Argimusco's area is characterized by the Capo d'Orlando Flysch siliciclastic sequence (ISPRA 2011), predominantly with the lithofacies CODb and CODa (Fig. 1). The basal conglomeratic facies CODa is represented by clasts, varying from few centimeters to 40–50 cm and sometimes even up to a meter, of metamorphic nature, granitoid, with rounded quartz grains and rarely limestones. These conglomeratic facies has been interpreted as a submarine escarpment deposit that precedes the deposition of sandstone turbidites CODb.

1.1 Plateau's Description

The area of the *Argimusco*, located in the province of Messina, on the island of Sicily, covers a wide plateau between 1165 and 1230 m above sea level, in the center of the so-called *Abacenino-Tindaritano territory*, where the harshness of Peloritani Mounts gives way to the sweetness of Nebrodi Mounts.

Situated near the town of Montalbano Elicona and the Natural Reserve of 'Bosco di Malabotta', the so-called *Argimusco Rocks* represent one of the rare examples of natural megalithic site of the whole southern Italy. It is the action of the Nature, mainly wind and water, which has shaped the huge rocks, creating stones from the particular human and animal figures. To describe the plateau and its main rocks I chose to use both an alphabetical nomenclature that some names given by historians of Montalbano Elicona and other nouns that I attributed personally (Fig. 2). Today the main entrance to the area area is located in the East part of the plateau,[2]

[1]The Age of these outcrop are: (a) COD (Chattiano-Lower Burdigalian, 23–20.4 Ma); (b) CFL (Upper Burdigalian-Langhian, 16–13.8 Ma); (c) ASI (Upper Cretaceous, 99.6–65.5 Ma).

[2]The Argimusco is situated along the SP115, called *Tripiciana*, to the junction that leads to *Malabotta's Wood*. Near this road junction there is the ruin of the so-called *Fondaco Tower*, which has a room with a barrel vault on the ground floor.

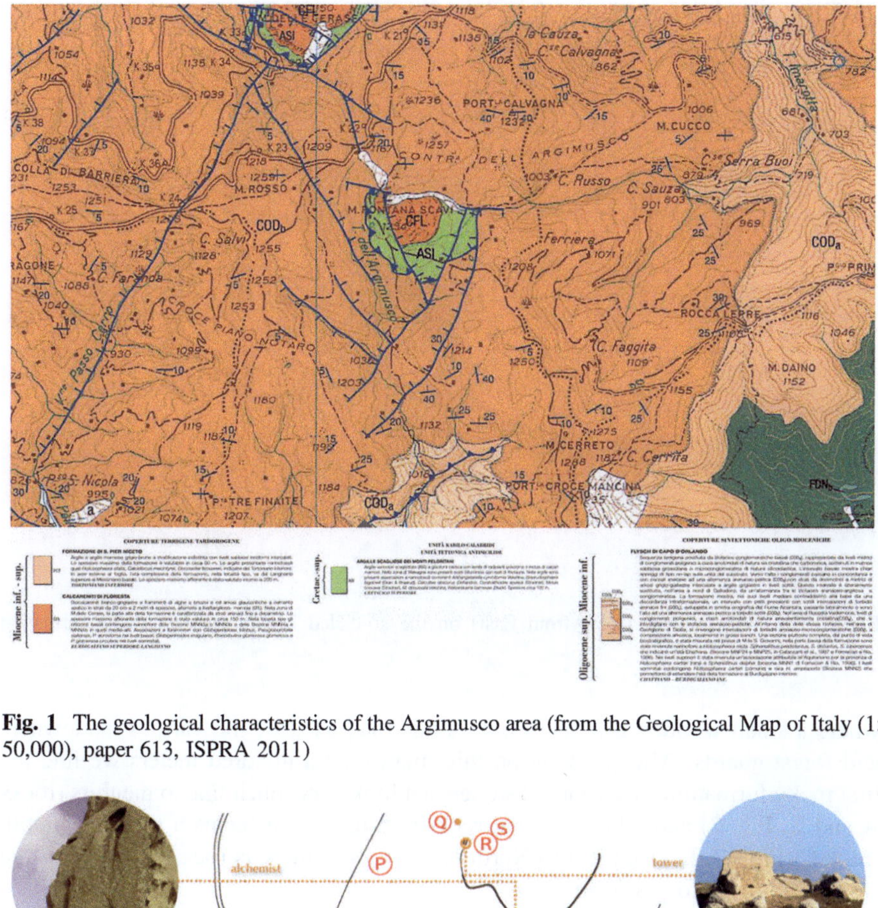

Fig. 1 The geological characteristics of the Argimusco area (from the Geological Map of Italy (1: 50,000), paper 613, ISPRA 2011)

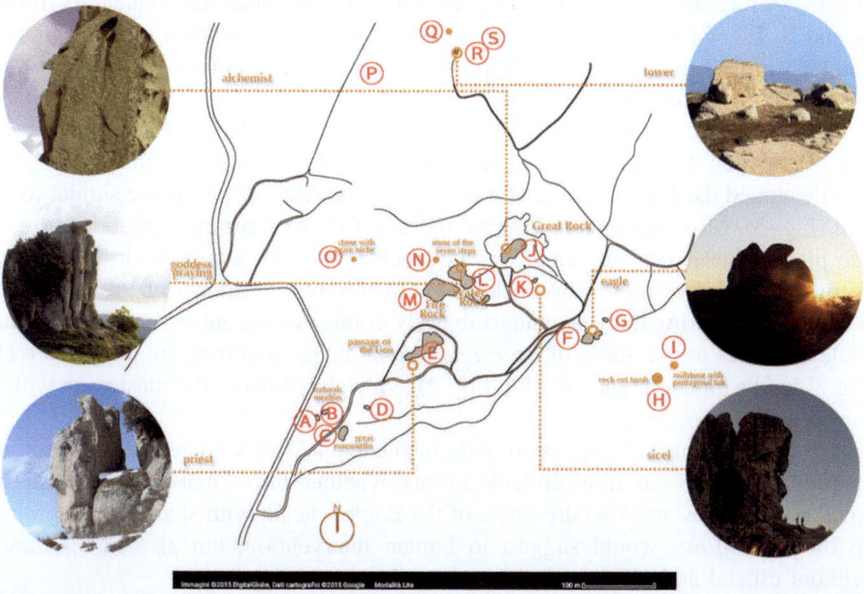

Fig. 2 The map of the Argimusco's Plateau with the position of the principal rocks

Fig. 3 The *Priest's rock* (view from East) on the so-called Lion's passage rocky formation (author's photo)

where a small wooden gate leads onto a dirt road used by the various landowners and forest guards. After walking on this small road a hundred meters we find the first rocky formations: two natural stones that look very much like to menhirs (rocks A and B, Fig. 2) and a large natural rock from the North seems a great mammoth (rock C, Fig. 2). Located just to North of the rock C there is rock D, known as the small mammoth for its form.

Continuing along the little road there is the so-called *Lion's Passage* (rock E, Fig. 2), a rock formation that in its southern part presents an anthropomorphic profile named the *Priest's rock*, a kind of human profile that has a hole similar to an eye (Fig. 3). Moving towards East we meet one of the most emblematic stones of the plateau, the so-called *Eagle's stone* (rock F, Fig. 2), a rock with the evident form of the bird of prey. This rock seems to observe the transit of the Sun in its apparent path during the year, and its majesty dominates the entire area. In the small valley that lies to the South of the *Eagle's stone* there are a rock-cut tomb (rock H, Fig. 2) and a millstone rock (rock I, Fig. 2), which has a particular master tub with a pentagonal shape.

On the western and the eastern side of the Eagle's rock there are two circular petroglyphs also, but it is difficult to say whether those marks are natural or man-made at present. Also the shape of the Eagle's head, with sharp beak pointing to the South-East, would suggest to human intervention, but also in this case, without official archaeological excavations, it's hard to say.

After reaching the central part of the Argimusco there is the so-called *sacred area*. It is definitely one of the most fascinating areas around the plateau, where we are immediately in contact with the grandeur of the *Great Rock* (rock J, Fig. 2), the largest rock formation and majestic of the plateau. On the eastern side of the Great Rock, if we look it from the south, we can see an impressive man profile, called the *Skull* or the *Sicel*. This profile has been clearly created by wind erosion processes, and in fact the entire large rock is characterized by alveolar cavities and streaks typical of the action of atmospheric agents. On the western part of the Great Rock is located another anthropomorphic profile: it is the so-called *Alchemist*. While in front of the Great Rock, to the south, there is another stone (rock K, Fig. 2) called *snakehead*.

From the sacred area we can admire the other 'big attraction' of the Argimusco: the so-called *Praying's stone*, a female figure with hands clasped whose profile, about 25 m high, that is located in the northern part of the so-called *Water's Rock* (rock L, Fig. 2). However also this anthropomorphic profile loses its suggestive shape when we get close to it; in fact placing it just below the high female figure can be seen the various cavities created by the wind over the millennia.

On top of the Water's Rock, just above the profile of the Praying's stone, we find one of the most important artifacts of the area: the so-called *Tub*. This cavity carved by the man in the sandstone, with rectangular dimensions of about 1.50 × 0.45 m, was defined by Gaetano Pantano a tub to collect water, which would serve to baptismal rites (Pantano 1994). Instead the scholar Giuseppe Todaro saw in this cavity even the tomb of the legendary *Arge*, personification of thunderbolt, the Cyclops son of *Uranus* and *Gaea*, the younger brother of *Bronte*, the thunder, and *Sterope*, the lightning (Todaro 1992). But in fact, closely watching the carving in the rock, we notice that the cavity does not have the typical cut away of the rocky overland tombs; also in the south-east angle of the cavity there is a drip, which suggests the function of a tank for collecting, probably linked to ancestral cults related to water.

Also on the Water's Rock, in the lower part of it, there is a sort of arch-shaped trench, a rock which is accessed via three steps carved into the stone, and from which we can have a quick and wonderful overview of all the plateau and the southern area, which culminates with the Etna. On this part of the rock formation are the trays for collecting water, and also in this case the presence of drips would seem to highlight the human manufacture of artifacts.

Just west of the Water's Rock meet the *Fire's Rock* (Rock M, Fig. 2), where there is a sort of inclined plane (approximately 30° with respect to the floor), on which we can see dozens of natural cupels, now consumed by water. This part of the Fire's Rock is called the *Sanctuary*. Proceeding in the description of the Argimusco plateau meet a stone that the scholar Todaro called the *Tetrahedron*, structure that I prefer to call the *Stone of the seven steps* (rock N, Fig. 2). The rock is characterized by the presence of seven steps carved into the sandstone that allow to arrive at its top. This stone is naturally projected toward the western horizon, and therefore it represents an ideal place from which to admire beautiful sunsets.

From the Stone of seven steps begins a small path, barely outlined among the ferns that cover much of the plateau, leading to the West. Along this little path, skirting the small grove of chestnut and oak trees that are located just west of Fire's Rock, we come to a big granite monolith; still continuing along the path we arrive to a rock that presents along the south side a shrine with the top shaped like a wishbone. This article is very interesting because it shows that in the Middle Ages the area was most likely placed in a kind of sacred journey or pilgrimage, then it seems to be linked to the Christianization of the place (Arlotta 2005). As often happens in areas where in the past were held ancient rites of pagan matrix, Christians have subsequently made several symbols of Christ on the stones, such as the cross. And on a rock located at the western end of the Argimusco district there is a petroglyph in the shape of a crucifix.

But back to the description of the site, so if from the central plateau, the sacred area part, continue to north, we arrive in the most northern zone of the area, where there are other rocky sloping floors and a rock formation with spirelets that for its particular form is called the *dragon crest* (rock P, Fig. 2). Not far from this rock there is the stone called *seal* (rock Q, Fig. 2). But the most interesting rock of the northern part of the plateau is certainly one that has a characteristic cubic shape and is called *Tower* (rock R, Fig. 2). This large stone lies right on the equinoctial (East-West) line; line passing through another large and important natural Rock, called *Rocca Novara* or *Salvatesta,*[3] that stands out with its distinctive triangular shape on the eastern horizon.

Complete the description of the plateau the area at East of the sacred area, where are the rocks of *Portella Calvagna* and finally *Monte Cucco.*

1.2 Toponym, Cartography and History

The etymology of the toponym is related to the first historical sources that speak of the Argimusco, which amounted to the Middle Ages. The first written mention dates back to 1282 and is in the form "Argimustus". This name appears in the *Historia Sicula* of Bartolomeo di Neocastro (Paladino 1921), the Messinese jurist who, narrating the facts related to the Vespers, describes the way of Peter III of Aragon, first king of Sicily (1282–1285), to reach Messina,[4] at the time besieged by

[3]This Rock is also known as the *Matterhorn of Sicily* and it is a Site of Community Importance (SCI). From geological point of view the Rocca Novara is characterized by its peculiar tectonic units (NOV, Cretaceous-Eocene age).

[4]*Bartolomeo di Neocastro*, *Idrisi* and *Malaterra* ensure the historical significance of the route, which goes from Randazzo leading to Messina and vice versa, in the Middle Ages; it is likely that the Nebrodi pass, near Montalbano and Argimusco, was already well known in ancient times, as is shown by an important literary passage from *Diodorus of Sicily* in relation to arms campaigns of Hiero II, which, in 271 BC, come to Tindari going up the river Flascio from Maniace and passes through Abaceno (Tripi) (Diodorus 22, 13, 2). These road links to the mid-twelfth century are

the Angevins. In this way the Catalan king, coming from Randazzo, passes near the locality "*Argimustus*":

> *Post haec ex parte illa jussit iter assumi, et dum pervenissent ad locum, qui dicitur Argimustus, jam Melatium, sicut in mare protenditur, insulae Vulcani, Lipariae et Strongylis ardentes conspiciuntur ex altis. Jam montium Phariae monstrantur confinia; satis visa placent, et loca commendnas delectabilia circumspectat; sedes Helenes Tindareae, ubi Virginis hodie sacra domus excolitur, Pactas et quae ante oculos surgunt Castra commendat; et descendens apud Furnarum, ibi residens noctem fecit.*

From the description we learn that the place was appreciated for its scenic quality already at that time, a magnificent overview on the Tyrrhenian coast, with the magnificent Aeolian Islands, on the promontory of Milazzo, on Tindari and the surrounding hillsides. A few years later, in July 1308, there was a first significant variant of the etymon. The letter of King Frederick III to his brother James of Aragon (Fig. 4), in which the sovereign answers to a proposed truce towards Robert of Anjou, Duke of Calabria, is released from the same place in which it was passed the father and is now called "*Argimuscum*".

After a couple of centuries, in the sixteenth century, two famous Sicilian scholars, the mathematician Francesco Maurolico and the historian Tommaso Fazello, does not carry any toponym of the plateau in their works depicting the island of Sicily,[5] while the Benedictine abbot and royal historian Vito Amico, in 1760, in his *Lexicon topographicum*, refer it in *Arcimusa/Archimusa* form.

As it regards the toponyms is very interesting to consider the work of cartographers also, which since the sixteenth century have left us valuable insights on the Argimusco's plateau. Among the first cartographic representations we find that of the Flemish mathematician and astronomer Gerhard Kremer, better known as *Gerardus Mercator*, that in 1589, in his famous *Regnum Siciliae*[6] map does not specifically indicate the place but report the presence of a water source on the

(Footnote 4 continued)

recorded as early as Idrisi, which registers between Randazzo and Montalbano a distance of about 20 miles (Idrisi 2008), and, indirectly, in 1061 AD by Geoffrey Malaterra in connection with the Nebrodi crossing wanted by Robert Guiscard and Count Roger, who <<... *audaciores sub Scalatripoli (today Tripi) hospitium sumunt. Inde in crastinum to Fraxinos perveniunt, et a Fraxinis to Maniaci pratum...*>> (Malaterra 2000), so along the road from Tripi to Montalbano and passing to Argimusco towards Randazzo to finally arrive in Maniace.

[5]Maurolico, F., *Descrizione dell'Isola di Sicilia*, Venezia, 1546; Fazello, T., *De Rebus Siculis decades duae*, Palermo, 1558. For the purpose of the study it is interesting to note that in kit of the Maurolico's opera there was also a map of the island, engraved in wood and printed the year before by the same printer in Venice and included by Giacomo Gastaldi in his famous cartographic collection dedicated to Italy; the design of this map is attributed by some to the same Maurolico (Samperi 1644). The Gastaldi map of Sicily is a real milestone in the history of the Sicilian cartography, as for the first time, following evidently to thorough critical review of the Ptolemaic paper, occur the merger between empirical and scholarly geography, and provides the main island's toponyms, both coastal and interior. This map will inspire the later works of Mercator and Ortelius.

[6]This map is now kept and exhibited at the National Library of France.

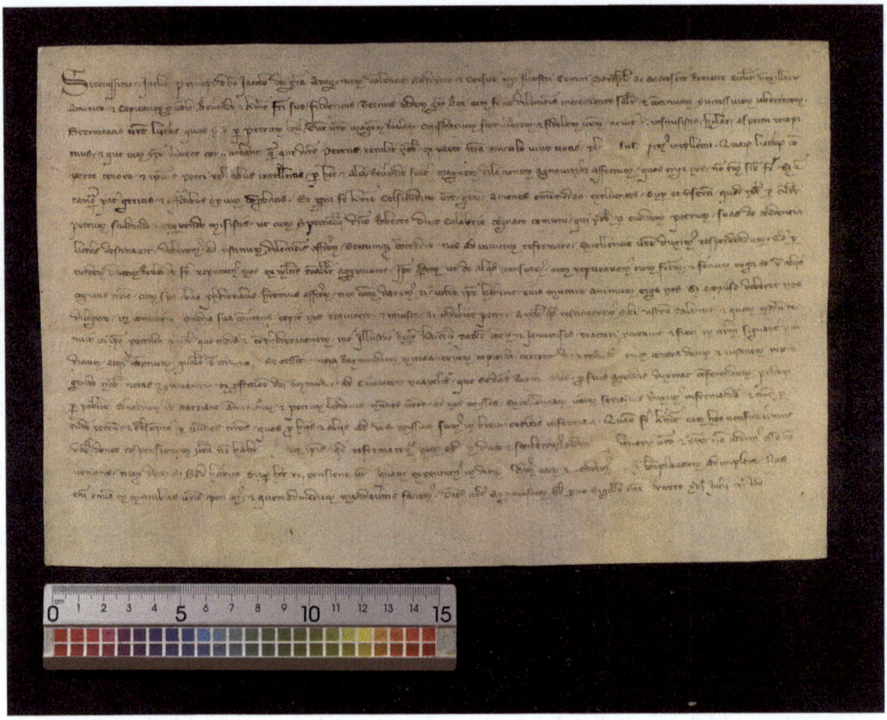

Fig. 4 The letter of King Frederick III was released from Argimusco in 1308 (Arxiu de la Corona d'Aragó, perg. 9944)

plateau, called *Lagrimusco fons* (Fig. 5), name then faithfully reported by many later cartographers, demonstrating an indisputable importance of the site as well topographic reference point. It has continued to report toponym and its water source in many maps drawn up between the seventeenth and the first half of the nineteenth century.[7] It must also be noted that in various other maps the water source water on the plateau, while not being reported with the name, is indicated with the typical graphic sign of water sources (see for example the Sicily's maps by *Nicolas de Fer* (1701 and 1722)).

As regards the meaning of the Argimusco's toponym, as recently evidenced by the historic Pantano,[8] it seems to have an Arabic origin, being a linguistic relic formed by terms "hağar" (to read *asgiàr*), that is 'stone, rock', and "misṭah", which means 'flat area', whose union derives "hagar-mistah", which translated

[7]I have identified 38 geographical maps of Sicily with toponym refers to the plateau, including those of (to view all the maps see the work of Dufour and La Gumina 1998), Quad von Kinckelbach (1600), Metellus (1601), Merian (1635), Mercator (1630), Jansson (1640), Blaeu (1640), Duval (1672), Collignon (1676), Wit (1680), Jaillot (1681), Bulifon (1692), Funcke (1693), Coronelli (1696) and Visscher (1698).

[8]Pantano G., *Argimusco, la verità sulll'origine del nome*, Centonove, Messina 2015.

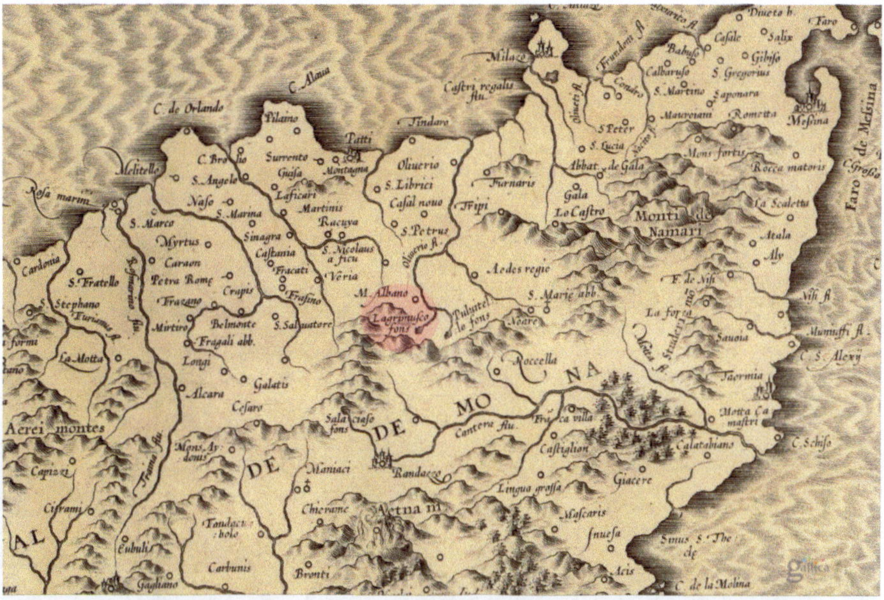

Fig. 5 Parte of the Mercator's map of Sicily with the reporting of the toponym in the form *Lagrimusco fons* (elaboration from *Regnum Siciliae*, Mercator 1589)

means '*Plateau of the Rocks*'. Semantics that corresponds perfectly to the physical and orographic characteristics of the place, in complete tuning with what is reported by Messina reporter Bartolomeo di Neocastro in the thirteenth century. In a more specific manner, "asgiàr" switch for metathesis to "argi" and "mistah" becomes "mustuh", from which "*argi-mustu*" that, latinized, becomes the "*argimustus*" of Neocastro, which, ultimately, adapts to its Latin language the Arabic terms that meet and that are reported to him in this place. So when in the last days of September 1282, the Neocastro stops with King Peter and his troops in this spectacular "Plateau of the Rocks" (Fig. 6), not knowing the name of the locality, he is forced precisely to ask, as he demonstrates with his eloquent expression "qui dicitur argimustus", that is 'which call Argimustus'. The reference was clearly to the local followers of the Muslim populations that a couple of centuries earlier had given the name to this ineffable place.

In the XX century some scholars have proposed different theories about the megalithic plateau. In the book *'Montalbano Elicona. Storia e Attualità'* by Nicola Terranova (1986), the Argimusco is mentioned several times, especially in reference to the possible passage of the Roman legions under the command of Lucius Cornificius and Octavian Augustus (I century BC), at that time in fight against Sextus Pompey. In regard to these matters Terranova cites the works of Cicero (*In Verrem*) and Appian of Alexandria (*Appiani bellorum civilium liber quintus*). In this context it is interesting to mention the presence of a megalithic complex which is located in Portella Zilla, about 4 km west from Argimusco, which may have been

Fig. 6 Peter III of Aragon on the Pyrenees; this picture shows a very similar representation of the rocky landscape of the Argimusco (Mariano Barbasán Lagueruela, oil painting on canvas, 1889)

used as a fortress during the Roman era, an architectural used to protect one of the mountain pass that led from Tindari to Taormina (Saporetti and Varisco 1999).

The hypothesis of Terranova, Saporetti e Varisco are supported by the fact that in this area passed right one of the so-called *regie trazzere*. It is one of the secondary roads that will certainly put in connection the Tyrrhenian coast with the Alcantara and Simeto Valley since antiquity. We know from the *Tabula Peutingeriana* (Fig. 7) the main ancient roads, the *consular roads*, which allow you to make the periplus of Sicily (Uggeri 2008), however in it is not indicated the secondary roads, many of which in Medieval Ages become the above-mentioned *regie trazzere*. Just in the Argimusco district flowed in the regia trazzera the roads from Roccella Valdemone, Malvagna, Tripi and San Basilio, one of the villages of Novara di Sicilia.

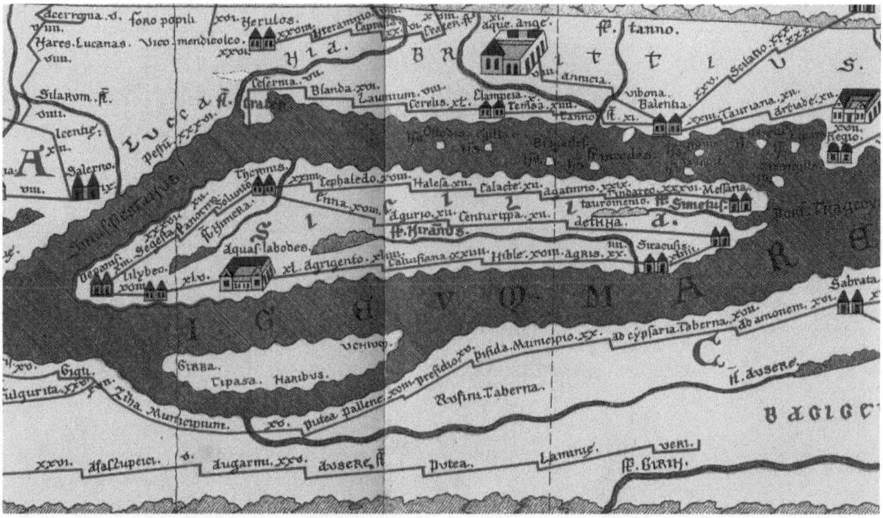

Fig. 7 Sicily with its consular roads represented in *Tabula Peutingeriana* (it is currently kept at the Austrian National Library in Vienna)

All that seems highlighted by several cartographic sources, such as, for example, the Smyth's map of 1826 (Fig. 8), in which right next to the Argimusco area is marked a road junction,[9] to the center of which stood a fondaco. The fondaco, to which were attached a now ruined tower and a tavern, was the customs office in the Middle Ages, as well as a resting place for travelers and probably a Roman stationem (Todaro 1992). Recently it has been suggested as most likely in the Montalbano territory arose a medieval *hospitalis* for pilgrims traveling to Jerusalem and Santiago de Compostela (Arlotta 2005), and it is interesting to observe that it was customary to build such hospitality structures right next to the ancient *stationes*.

2 Archaeology

It was never made any official archaeological dig on the Argimusco's plateau, but from surface surveys and archaeological finds is it possible to assume that the site was attended in prehistoric and protohistoric ages.

---

[9]Same observation can be done to another map, fundamental to the history of Sicily, the *Carte comparée de la Sicile du XIIe siècle*, drafted in 1859 by A.H. Dufour and M. Amari (Bibliotheque Nationale de France, GEC-2369) on the basis of the *Book of Roger* written by Al-Idrisi in the eleventh century. Although the Arab geographer did not mention the Argimusco, the map of the Dufour-Amari, similar to that of Smith, put the place in an important crossroads leading: to the North-West towards Montalbano, West to Floresta, South to Roccella, North-East towards Novara of Sicily and East to Pizzo Bonavi. It is to note the disappearance of the water source 'Lagrimusco' from the nineteenth-century cartography, evidently counted within the Argimusco district.

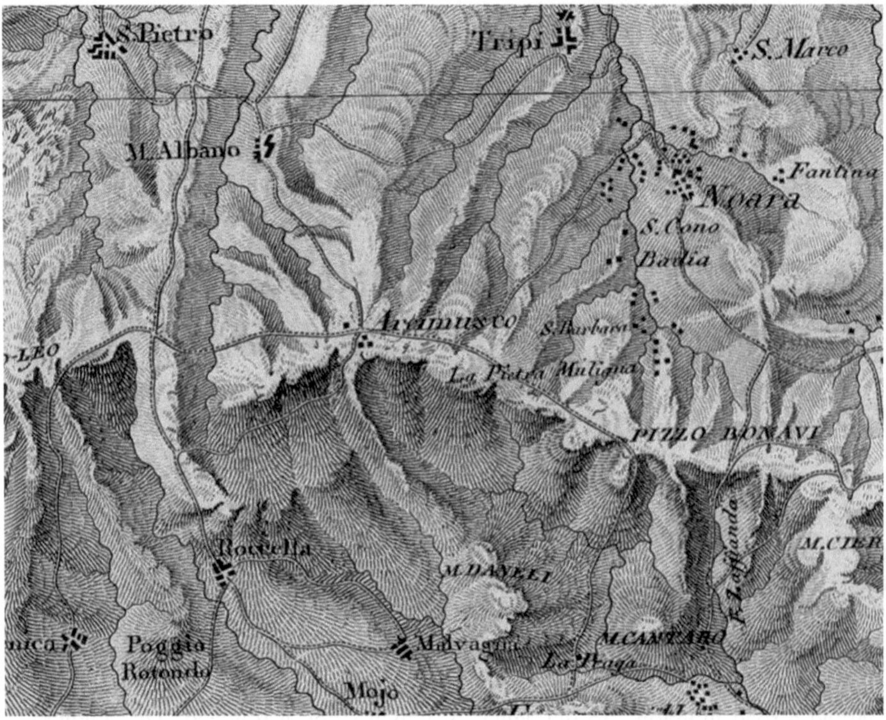

Fig. 8 The road junction of the Argimusco (*Arcimuso* in this map) district marked in the *Carta generale della isola di Sicilia* (1826) by G.E. Smyth (Bibliothèque Nationale de France, GED-4625)

The Argimusco in fact is at the center of an area that has been frequented by Sicilian ancestors over the millennia, indigenous peoples such as the Sicanians and the Sicels who have left indelible traces of their allocation in Sicily (Procelli Albanesi 2003; Ciaceri 2008). In an easterly direction, a few kilometers from Argimusco, there is, for example, one of the two Mesolithic shelters discovered so far in Sicily: the *Sperlinga of San Basilio* (at Novara di Sicilia), a rock shelter that has been utilized up to the Bronze Age (Brea 1958; Cavalier 1971, 2000). In the same direction there is also the *Rocca Salvatesta* of Novara di Sicilia, where there is a settlement of perhaps Late Antiquity or Early Middle Ages ever dug and some rock shelters probably Neolithic (Cavalier 1971).

While further North-East there are the seasonal settlement of the early Bronze Age of *Monte Bammina* (Martinelli and Prosdocimi 2009) and the important necropoleis of *Longane* (e.g.: Brea 1967; Martinelli 2000a; Campagna 2003), where it was found the famous caduceus bronze now housed in the British Museum. Always in the North-East direction, near the town of Tripi, lies the Sicel-Greek necropolis of *Abakainon*, a hellenized indigenous center of which still have to be found the urban remains (Bacci and Coppolino 2009; Sofia 2015; Sofia 2016).

Fig. 9 The Rocca di San Marco, the archaeological site located between the villages of Floresta and Ucria along the SS116 (author's photo)

To the North there is the site of *Tindari*, and few know that inside the archaeological area, along with the famous Greek-Roman ruins, there is also a little-known prehistoric site (Cavalier 1970; Martinelli 2000b; 2005). To the West there are the *Rocca di San Marco* (Fig. 9), a rock shelter of Paleolithic (Brea 1965; Martinelli 2000c) and the megalithic complex of *Portella Zilla* (Saporetti and Varisco 1999). Moreover to the South of Argimusco, in the northern part of the Alcantara Valley, is located *Rocca Pizzicata*, a rock settlement that has oriented altars carved in the sandstone, some millstones rocks and a rock cut tomb (Platania and Scaravilli 2013; Orlando et al. 2016).

We must also remember that in the whole Alcantara Valley there are many important archaeological sites, such as: Naxos, Petra Perciata, Francavilla di Sicilia, Randazzo e Sciare di Santa Venera (for example: Lentini 2006; Spigo et al. 2008; Orlando 2015; Orlando et al. 2016).

In the area around the Argimusco's plateau there are also artifacts ranging from the Prehistory to the Middle Ages (Todaro 1992); but unfortunately this area has not been the subject of a workforce plan for the use, protection and archaeological study, and therefore many theories are still waiting for a valid scientific evidence.

From the above considerations it is more than evident, therefore, that the Argimusco plateau was to be almost a necessary step for the ancients who moved seasonally to hunt among the lush forests of the Nebrodi and Peloritani, and those from the coast Tyrrhenian moved south to the Etna area, or vice versa, using the zone crossings (Portella Zilla, Portella Mattinata, Portella Mandrazzi). Traces of such attendance can be found even today in the same rocks of the plateau, where man has left indelible signs of its activity.

Fig. 10 The stone with the rock-cut tomb present in the small valley to the south of the plateau (author's photo)

2.1 The Rock-Cut Tomb and the Rocky Millstone

The anthropization of the Argimusco site is more than evident, for example, if we look at the rock-cut tomb (Fig. 10), the particular rocky millstone,[10] the stone of the seven steps, the tub and little basins present on the Water's Rock. The rock-cut tomb and rocky millstone lie just south of the Eagle's rock, where there is a small valley (*Fosso Sauza-Fontana Schiavi*). Both monuments were made using two large sandstone natural stone, in a time not yet identified.

The rock-cut tomb (2 m length, 70 cm height) is characterized by two lines engraved in the stone, placed over the entrance of the cavity, that allowed to drain rainwater. The rocky millstone, which is a few meters away from the tomb, is characterized by a main tub with a pentagonal shape, currently unique in Sicily (Fig. 11). And it is this artifact that could allow us to make new and interesting discoveries on the plateau dell'Argimusco. For some years in fact Gloria Olcese of the 'La Sapienza' University in Rome has launched a census of rocky millstones in Sicily, a research project aimed at the study and enhancement of these fascinating artifacts that man has used since ancient times to produce the wine. The importance and the great novelty of this study is that through the morphology of the basins seems to be able to date the millstone so that the site.

[10]The rocky millstones are generally formed from two adjacent tubs, different in shape and size. The largest is located at the top of the stone. A hole dug into the rock connects the two tubs. In the upper tub was crushed grapes. After crushing the grape must was collected in lower tub.

Fig. 11 The rocky millstone with the main tub with pentagonal shape highlighted (photo and elaboration by author)

The group of Roman scholars has so far analyzed only some areas of Sicily, including the Alcantara Valley, where there are dozens of rocky millstones (Puglisi 2009). But also in the abacenino-tindaritano territory there are dozens of millstones, just think only to the many artifacts in the villages of Tripi and Basicò. We know that with the advent of the Greeks in Sicily vine cultivation and wine production had a great development (for example: Brun 2003a, b); in this regard, just remember the iconography in Naxos coins (Fig. 12). The millstone with pentagonal tub of the Argimusco may be of earlier age of the arrival of the Greeks, moreover paleo-botanical surveys have already confirmed that Sicel produced wine in Sicily as early as the tenth century BC exploiting the wild vine.[11]

2.2 *The Water's Rock*

The so-called Water's Rock, located in the central part of the Argimusco (rock formation indicated by the letter C in Fig. 2), is probably one of the most important places of the plateau, and its careful analysis allows us to propose a use of the area as a sacred site, where they carried atavistic rituals dedicated to the worship of water. Access to the Rock is to the West, some natural steps allow to reach a first

[11]On the topic of wine production in antiquity there are numerous works; for example, as regards the western Sicily, I want to remember: Costantini L., *Semi e carboni del Mesolitico e Neolitico della Grotta dell'Uzzo*, Quaternaria, vol. 23, 1982; while as it regards the eastern Sicily: Amato F., *Prospettive di ricerca sulla produzione vitivinicola antica a Licata (Agrigento)*, 2012.

Fig. 12 Up: silver drachma (O/Head of Dionysos to left, with long, pointed beard, ivy wreath in his hair and a plain torc-like necklace bordered by dots at the truncation; around, border of dots within two linear circles. R/Bunch of grapes on stalk with two leaves. Around the figure, the inscription: ΝΑΞΙΟΝ (of Naxians)) (5.57 g), Naxos, Sicily, 530–510 BC; down: tetradramm (O/Head of Dionysos to right, with long wavy beard, hair tied in a knot at the nape, short bushing front of the ear. R/Silenus sat on the rock with kantharos. Around the figure, the inscription: ΝΑΞΙΟΝ (of Naxians)) (17, 19 g), Naxos, Sicily, 461–430 BC

level, where there is a rocky platform, which could be defined as a kind of *altar*, which is accessed via three steps carved in the sandstone. The altar is oriented towards the South, on it there is a walkway/artificial corridor (length 13 m, width of about 0.60 m, depth 0.35 m) and some elliptical trays to collect water, which present a spillage point for the water that is generally located to the east (Fig. 13). This platform has a natural extension to the east, pointing toward the Rocca Salvatesta, overlooking the medieval town of Novara di Sicilia.

From the first level, that of the rocky platform, we can move to the highest level, where there is an artificial rocky tub (length about 1.80 m, width of about 0.90 m, depth 0.50 m) and another couple of little tubs with gutter for water drainage (Fig. 14). It is very interesting to note how these little tubs are similar in shape and size to those found on *Monte Croccia* in Basilicata, in the archaeological site of *Petra de la Mola* (Curti et al. 2009) (Fig. 15). The great tub and the system of tubs for the collection of water allow to propose the valence of the site as a worship area. The water certainly was not lacking on the plateau, and different water sources still today make lush the entire area, which is full of ferns, plants and wild herbs. The worship of water is one of the oldest in Sicily and throughout the Mediterranean, and on this issue there is a fair literature (for example: Bianco 1999; Melis 2008; Portale 2012).

Fig. 13 The so-called *altar* on Water's Rock (photo credit Massimo Calcagno)

Fig. 14 The tub carved on the Water's Rock (photo credit Massimo Calcagno)

So we have to assume that water played an important role for the Argimusco, and this, as already mentioned, it is also testified by the numerous cartographic maps showing the toponym linked to water sources. The Argimusco plateau would constitute so like a real *natural sanctuary*, a place used by people since ancient times to worship the Nature and ancestral deities. However, it is appropriate to reaffirm once again, until today has not been made any official archaeological excavation that might corroborate this hypothesis. However, not infrequently,

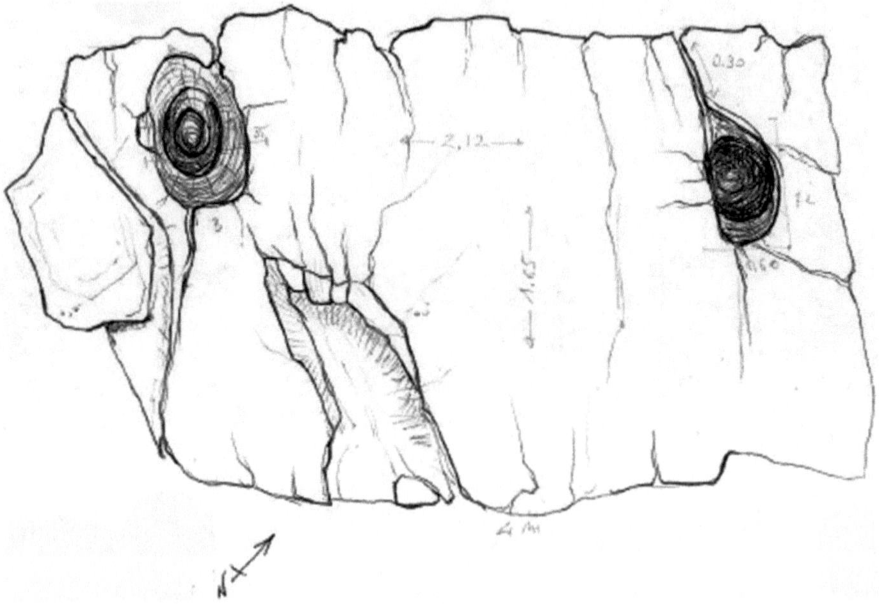

Fig. 15 Relief of a rock with little tubs present in the archaeological site of Petra de la Mola on Mount Croccia (Oliveto Lucano (MT), Basilicata)

walking on the plateau, is possible to observe archaeological remains consist of fragments of lithic and ceramic industries ranging from prehistory to the Middle Ages. An official excavation by the Superintendence of Cultural Heritage of Messina is therefore highly desirable, in order to determine once and for all the chronology of the frequentation of the site.[12] It is interesting to note that several recent archaeoastronomy's studies in Sicily have often emphasized the correlation between the worship of water and sky observation (for example: Nigro 2010; Orlando 2016; Gori and Orlando 2016; Hannah et al. 2016a, b; Hannah et al., in this volume; Orlando and Gori, in this volume), and also to the Argimusco, as we shall see shortly, it seems possible to suggest that in-teresting combination.

3 Astronomy at Naked Eyes: Introduction

We know that man since Neolithic times, unemployed from the life of hunting and gathering, began to create a calendar based essentially on the sunrise. The need to establish a calendar, useful both for farming and for purposes of worship,

[12]For this purpose I am working to activate the *Argimusco Project*, forming an international group of research that in 2018 should start the multidisciplinary study on the plateau.

became more pressing when the man found sustenance in the use of agriculture and animals husbandry (Hoskin 2009).

If we observe from the same place, for instance, dawn during the course of the year, we may notice that the Sun moves from day to day. In particular, there are four sunrises during the year that can be used as references for a handy horizon's calendar, those related to the two solstices and the two equinoxes. It then creates three points on the horizon, that mark the seasons and therefore the periods in which to sow, harvest or proceed to certain agricultural work. If we imagine to follow the apparent motion of the Sun from the winter solstice, our star will rise to a point at south-east in that day, describing an arc in the sky very low on the southern horizon. Subsequently, the point of sunrise will shift to the east, and the arc described will increase. At one point, on the day of the vernal equinox, the Sun will rise exactly in the east. Then the point of sunrise will move to the North-East, where it reaches an extreme limit on the summer solstice'day, the moment in which the arc described from the Sun will reach its maximum (Fig. 16). From this moment onwards the motion of the point of sunrise is reversed, starting to shift to the right (South). The Sun will rise again in the east to the autumnal equinox, and then, continuing its southward shift, day after day, the point of sunrise returns to rise in the extreme South-East. At this point exactly one year has elapsed. Starting from one of the two extremes and, for example, counting the number of days, is possible to determine the duration of the year in a simple way.

On the world, from prehistory to the present day, there are countless examples of horizon's calendars (Fig. 17), exploiting natural contours of mountains (e.g.: MacKie 1977) or particular totem (lytic or wooden), suitably placed, which are been used to highlight points on the horizons (Aveni 1982). These devices have been used to create practical calendars (for example: Zuidema 1964; Bauer 1998; Ghezzi and Ruggles 2007). Among them, the calendars that have been extensively studied, and used up to now, there are Hopi's calendars (McClusky 1982).

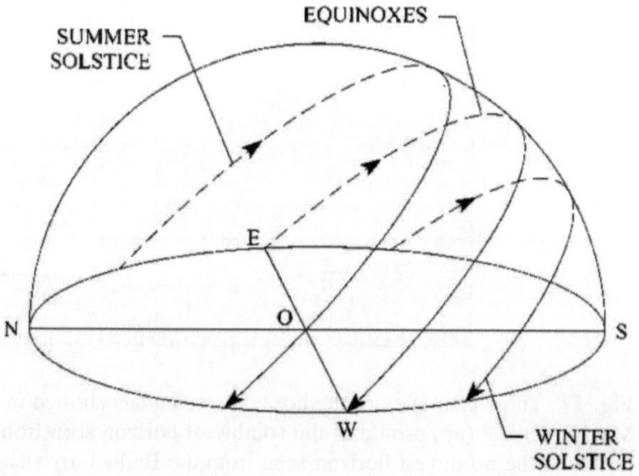

Fig. 16 The path of the sun during the year, viewed by an observer at a latitude above the tropic, with a flat horizon

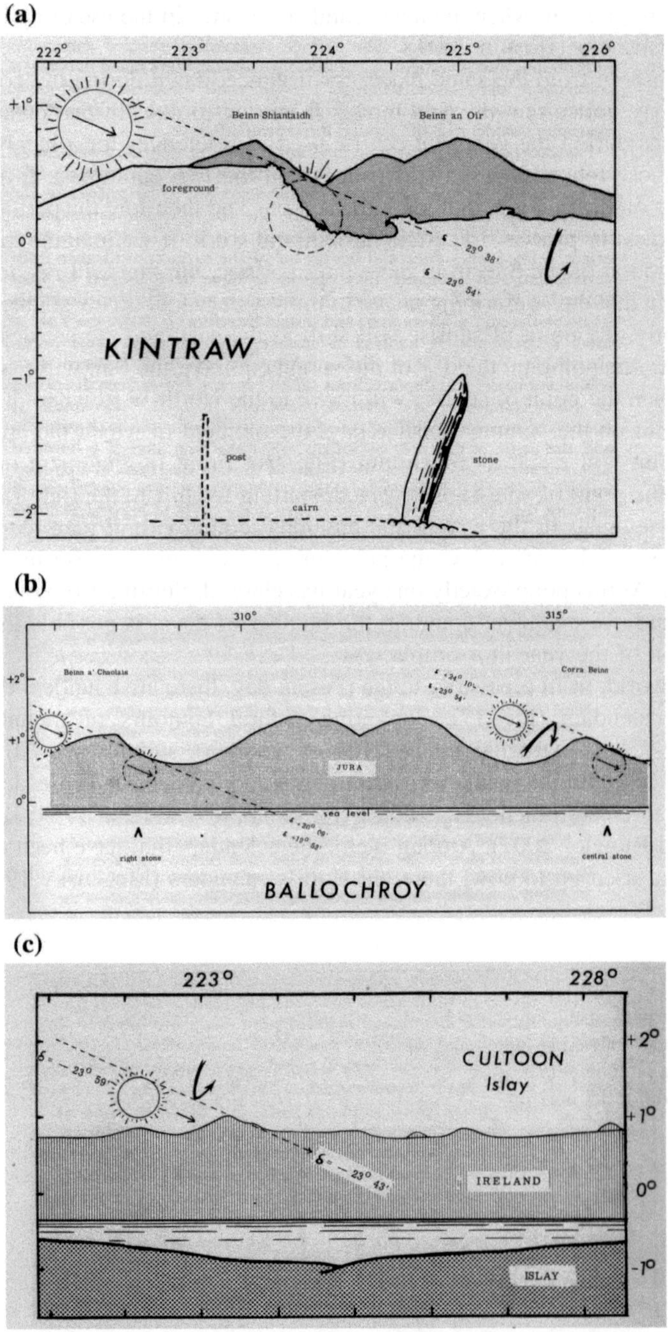

Fig. 17 Three examples of the horizon's calendars studied in Scottish archaeological sites (from MacKie 1977): (up) profile of the southwest horizon seen from the platform at Kintraw; (middle) profile of the northwest horizon seen from the Ballochroy standing stone, Kintyre; (down) profile of the southwest horizon seen from Cultoon

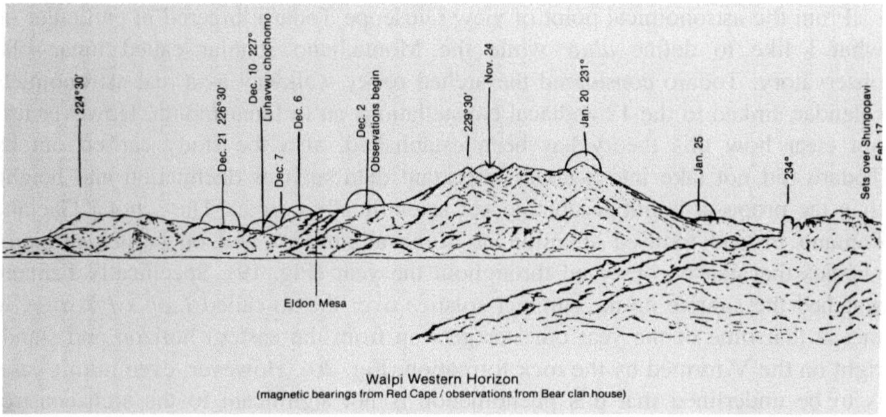

Fig. 18 One of the horizon's calendar of the Hopi, designed by A.M. Stephen while he was in the Walpi village (First Mesa, Arizona) at the end of XIX century (from McClusky 1982)

These horizon's calendars were analyzed by the anthropologist Alexander M. Stephen, who lived among the Hopi in the last decade of the XIX century. The horizon's calendar created by the village Walpi (Fig. 18), for example, shows the observations of a priest of the Sun on the eve of the winter solstice, in order to fix the date of commencement of the ceremonies (*Soyal*). The sunset of the Sun in the crook reached on december 10th (between the peaks of the San Francisco Mountains, near Flagstaff), provides the signal that 4 days after the nine-day celebrations of the winter solstice had to begin (Stephen 1936).

3.1 *The Studies of Todaro and Pantano's Scholars in XX Century*

Among the first scholars who have carried out research on the Argimusco's plateau and the abacenino-tindaritano area there are two professors of high school of Montalbano Elicona: Gaetano Maurizio Pantano and Giuseppe Todaro.[13] Fascinated by the innumerable ancient structures and rocks formations spread throughout the zone, interpreted as tangible signs of an ancient and unknown civilization of the past, the two researchers studied more than twenty years the whole area, discovering interesting prehistoric sites. Part of this research has been incorporated into two essays: *Alla ricerca di Abaceno* (Todaro 1992) and *Megaliti di Sicilia* (Pantano 1994). Of these studies is need to emphasize the far-sightedness of the proposal for a cultic and primitive valence of the Argimusco.

[13]Prof. Giuseppe Todaro has sadly passed away in January 2017.

From the astronomical point of view Giuseppe Todaro lingered in particular on what I like to define *altar* while the Montalbano scholar called lunar-solar observatory. Todaro considered the arched rocky walkway as a real astronomical calendar, linked to the 12 zodiacal constellations an to lunar month. However it is not clear how this theory has been established, also the study carried out by Todaro did not take into account important data such as declination and height, then the proposal is not acceptable nor scientifically correct. The scholar Gaetano Pantano instead focused attention on some alignments and on particular hierophanies that can be observed throughout the year (Fig. 19). Specifically Pantano watched the sunrise on the summer solstice over the so-called *Lion's Passage*; in fact at that time of the year our star gets up from the eastern horizon and stands right on the V formed by the rock formation (Fig. 20). However, even in this case, is to be underlined that this phenomenon is not significant to the archaeoastronomical level, since the choice of the place of observation is completely arbitrary, and not marked by a specific point on the ground, that is, a point where we should find a so-called marker, which could indicate a place of worship/observation. In a similar way is meaningless the fact of going to seek astronomical alignments at Eagle's neck in the days of the solstices, because even in this case the search is done by choosing the direction of observation in a totally arbitrary way, that is, I repeat, without having any point or structure on the ground that marks the possible viewing position. The position to view a celestial phenomenon on the horizon cannot choose in inductively way, and thus in speculative way, but must be stated in a clear and obvious way on the ground. Among the Pantano's studies deserves more attention the observation of solar dawn at the equinoxes from the Fondaco tower (Fig. 21). This alignment is interesting in that the construction of the medieval tower at the point where it is, and probably of the Roman stations also, could not have been chosen at random; this point is indeed significant as from that place we see the sun rise above the Lion's Passage during the equinoxes (spring and autumn). Unlike Montalbano scholars from several years instead my studies have focused on the observation of the horizons, practicing what is known as *astronomy horizons*.

3.2 Argimusco and the Astronomy's Horizons

At the Argimusco the horizons' profiles are particularly well suited to the observation of dawn and sunset of the Sun. Being on the plateau we have all horizons free, and the view spans at 360°, from north to south, following the so-called *volcanoes' line* (from Etna to Stromboli), and from east to west, along the so-called *light's line* (from sunrise to sunset). And still today, there is a horizon that most of the others provides important information for the calendar purposes: the eastern.

On this horizon is silhouetted infact the Rocca Novara (1340 masl), near the village of Novara di Sicilia, which serves now as in the past as an indicator of the equinoxes. There are several places on the plateau from which we can observe

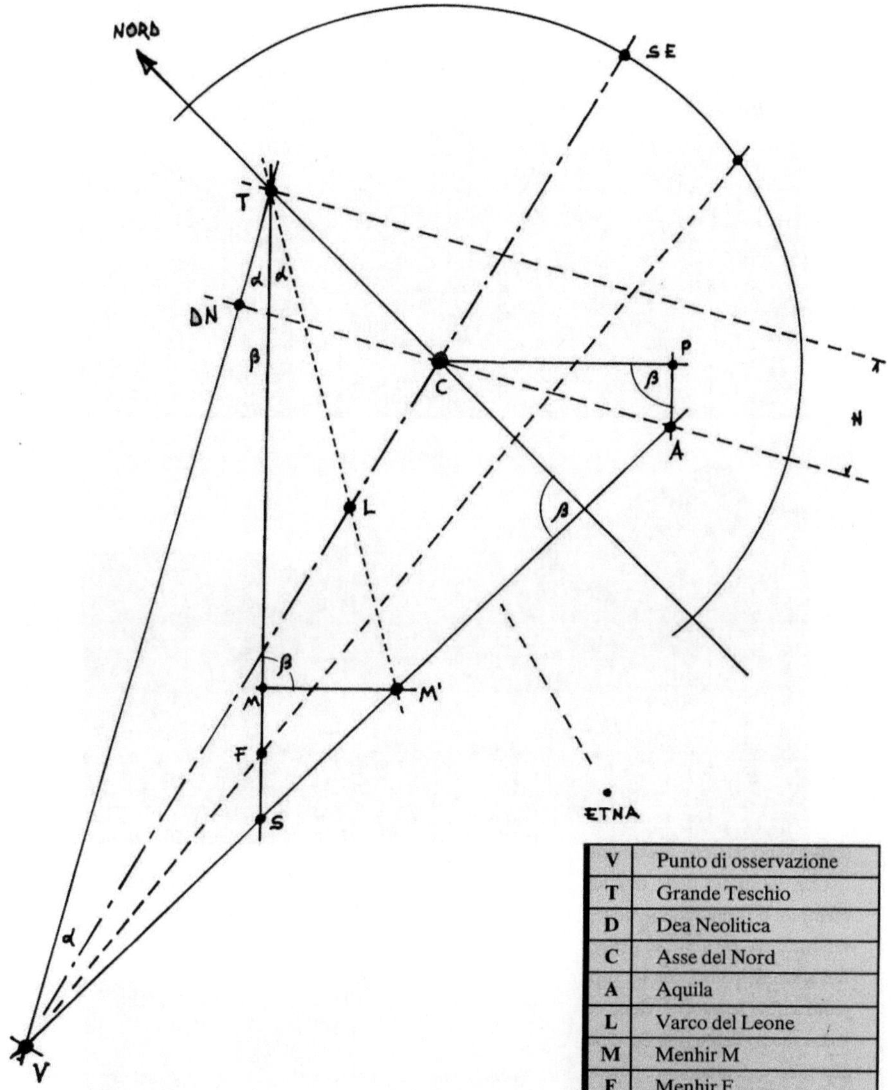

V	Punto di osservazione
T	Grande Teschio
D	Dea Neolitica
C	Asse del Nord
A	Aquila
L	Varco del Leone
M	Menhir M
F	Menhir F
M	Menhir
S	Sorgente
N	Area di necropoli
SE	Solstizio d'estate
P	Propiziazione Piogge

Fig. 19 Diagram made from Pantano for his theory of the alignments (from Pantano 1994)

Fig. 20 Sequence of the sunrise on the *Lion's Passage* in the summer solstice; the observation point is the hill opposite the current entrance (from Pantano 1994)

Sequenza che presenta il sole, mentre sembra fermarsi, sul varco del Leone il giorno dell'equinozio d'autunno (23 settembre)

Fig. 21 Sequence of the sunrise on the Lion's Passage at the equinoxes; the observation point is the Fondaco tower (from Pantano 1994) the eastern horizon, but only two have a probable value of privileged points of observation: the stone called *Tower*, in the northern area of the Argimusco, and the so-called *Altar* on Water's Rock. Just these two places have been used to carry out preliminary studies of archaeoastronomy (see next paragraph). We need to remember that the Tower rock is located just to the geographical East-West line.

Then watching the sunrise during the year from the Argimusco plateau, if the dawn of the sun is observed near Rocca Novara the period coincide with the equinoxes. If the Sun rise on the left side of the Rocca Novara (northeastward), immediately after the spring equinox, begins the period (spring season) leading to summer. While if the Sun rise to the right of the Rocca Novara (southeastward), immediately after the autumn equinox, begins the period (autumn season) leading to winter (Fig. 22). Thus making use of this simple horizon's calendar, the Sicilian ancestors, for instance, the Sicels of Abakainon's town, could know, with some precision, the seasons and create their own useful calendar for agricultural activities and worship. So the Argimusco and Rocca Novara are closely linked, together form what is often now defined a *sacred landscape*.

In recent years, some scholars (Devins and Musco 2014) have proposed improbable theories about the *Argimusco's stones*, including one that says *"... the megaliths of the Argimusco are in relation to the heavenly constellations, not in alignment as in other megalithic sites in the world, but as a reflection of the same constellations"*. It should be stressed that this theory does not find any scientific basis; I also feel that it is completely misleading to speak of megaliths, as the word *megalith* refers to a large stone or a set of stones used to build a structure or monument without use of binders such as lime or cement. On the Argimusco plateau, it is worth repeating, are instead only natural stones.

Instead I like to think that the sky observed in ancient from the Argimusco was able to inspire men to choose celestial asterisms (Fig. 23). Every civilization in the past chose their own constellations, and no one will stop to believe that even the people who lived in Sicily in ancient times, for example the Sicani, the Sicels and the Elymians, have had their own constellations.

3.3 Archaeoastronomy: Preliminary Analysis

The first archaeoastronomical reliefs on the Argimusco plateau began in 2012: with the use of compasses and inclinometers I studied the directions of different alignments; this research activity lasted about 2 years. In 2014 however, usingsatellite data (from Google Earth), I measured several azimuth, considering different points on the plateau where it was most evident human action (Orlando 2016). At this preliminary stage my studies have focused on the eastern horizon. As it has already been said, two places on the plateau seem to be very interesting to observe the eastern horizon, and in particular the solar dawn at the equinoxes: the so-called

Fig. 22 The Rocca Novara is still used today as an indicator of the equinoxes. The Sun rises on the left side of the Rocca Salvatesta, immediately after the spring equinox (up image), while the Sun rise to the right of the Rocca, immediately after the the autumn equinox (down image). The observations were made in this case by the Tower rock (photos and elaboration by author)

Altar on Water' Rock, and the so-called *Tower* in the most northern of plateau (Fig. 24).

In Spring 2015 it held the first campaign of field measurements (Fig. 25) and azimuth are measured with a high precision optical theodolite (Kern DKM2) with north calibrated using a long-distance GPS measure from the theodolite station to a recognizable feature (for example a corner of the Tindari Sanctuary); consistency of all measurements was cross-checked with compass-clinometer readings, corrected for magnetic declination and with Google Earth readings as well (due to various reasons, although the nominal accuracy of the instrument is below 1′, we estimate that the error of our measurements can be reasonably assumed to be ±15′). Data are reported in Table 1.

From the data shown in Table 1 it is quite clear that the point where we can better observe the hierophany of the 'Sun on the Rock' is the stone of the Tower. However while there are clear signs of human activity on the Water's Rock, and in

Fig. 23 A fantastic representation of a man in Argimusco who observes the constellation of the *Griffon* (image credit Lelio Bonaccorso)

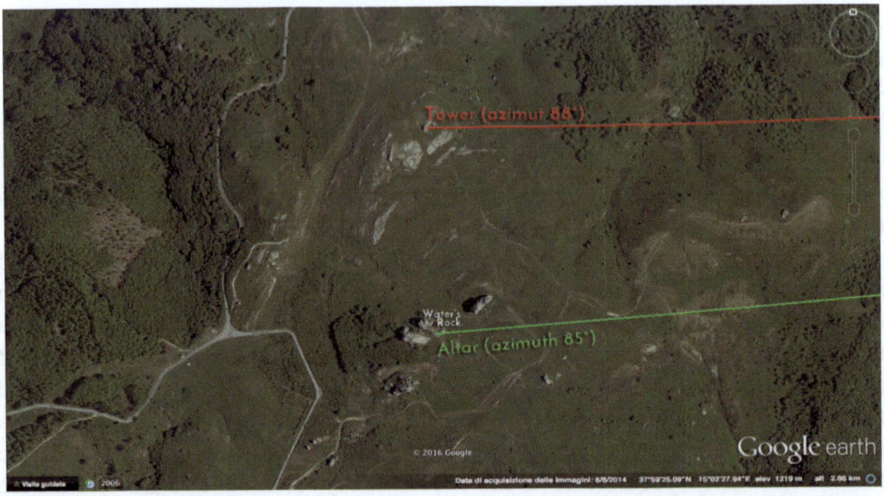

Fig. 24 Azimuth of Argimusco-Rocca Novara's system calculated with Google Earth (GE) considering the *Altar* and the *Tower* (elaboration from GE by the author)

Fig. 25 On Water'Rock during measurements with theodolite (author's photo)

Table 1 Data of azimuth, horizon, altitude and declination measured respectively from the *Altar* on the Water's Rock and from the *Tower* in the northern part of Argimusco plateau

	Azimuth (°)	Horizon (°)	Altitude (masl)	Declination (°)
Altar	85.00	0	1231	+3.90
Tower	88.00	0	1232	+1.7

particular on the so-called *Altar*, were not found signs of human settlement on the stone of the *Tower* at the current time. So it will be crucial to make archaeological excavations even in the northern part of the plateau.

4 Conclusions

In this paper I had the honor to present, for the first time at a scientific conference in Italy, the naturalistic and archaeological site of the Argimusco, a plateau that is located not far from the Tyrrhenian coast of Sicily, in the province of Messina. The Argimusco plateau has been known by the general public only in recent years, thanks to the many cultural events that I realized from 2012.[14] The plateau is now fairly well known by hikers, while little or not known in academic circles, both nationally and internationally. The Argimusco plateau is characterized by rocks which have anthropomorphic and zoomorphic figures, some of which have obvious signs of human activity. As it has never been made any archaeological excavation is rather difficult to determine the frequentation of the site based exclusively on the art rock.

The plateau is mentioned for the first time in some writings of the Late Middle Ages, during the Aragonese period, and then the toponym, linked to some water sources, it is found in several cartographic maps.

The Argimusco plateau is located in an area very rich in archaeological sites, both prehistoric to medieval, so the site is configured as a natural transit hub for travel between the ridges of the Nebrodi and Peloritani. The first amateur studies on the plateau were conducted by some historians of Montalbano Elicona in the second half of the twentieth century. Part of these studies has been dedicated to the observation of possible astronomical alignments.

My research on the Argimusco plateau has focused on the study of the rocks that have an obvious human activity and on the observation of the horizons (horizons' astronomy). In particular the eastern horizon's profile, dominated by the peak of the Novara Rock, it sill used today as an indicator of the equinoxes. An observer who watched the dawn of the Sun from the Argimusco during the year, could have clear indications of differents seasons, and useful informations about how to make a practical calendar. Two points are used for this study: the so-called Water's Rock and Tower.

[14]To watch some official videos of the festivals I propose a selection from the web, for example: https://www.youtube.com/watch?v=6AGUJioKw8A&t=4s ('*Pietre&Stelle2014*'); https://www.youtube.com/watch?v=IHX6FBlXdRo ('*Alla Ricerca dell'Astronomia Perduta 2015*'); https://www.youtube.com/watch?v=-JpBkOqt-a8&t=6s ('*Alla Ricerca dell'Astronomia Perduta 2016*'); https://www.youtube.com/watch?v=26YJXO62Ct0 ('*Pietre&Stelle2016*').

In order to complete the multi-disciplinary study of the area I proposed to activate the *Argimusco Project*, so that a group of international research could finally provide useful information to decipher one of the most fascinating and unknown archaeological site of Sicily.

Acknowledgements I thank very much some members of the Istituto di Archeoastronomia Siciliana (IAS), and in particular: Dr. Davide Gori, vice president of the IAS, for the geological contribution, and Dr. Alessandra Alberici for her extraordinary presence in the research activity on the abacenino-tindaritano territory. And finally I thank Prof. Gaetano Pantano, Prof. Giuseppe Todaro, Dr. Sergi Grau Torres and Dr. Gino Sofia, director of the Archaeological Museum 'S. Furnari' of Tripi (ME), for their collaboration and friendship.

References

Albanese Procelli, R.M.: Sicani, Siculi, Elimi. Longanesi, Milano (2003)

Arlotta, G.: Vie Francigene, hospitalia e toponimi carolingi nella Sicilia medievale. In: Oldoni, M. (ed.), Tra Roma e Gerusalemme nel Medioevo. Paesaggi umani ed ambientali del pellegrinaggio meridionale. Laveglia Editore, Salerno (2005)

Amato, F.: Prospettive di ricerca sulla produzione vitivinicola antica a Licata (Agrigento). In: Ciacci, A., Zifferero, A. (eds.), Archeologia della vite in Toscana e nel Lazio: dalle tecniche dell'indagine archeologica alle prospettive della biologia molecolare, Firenze, pp. 307–348 (2012)

Aveni, A.F. (ed.): Archaeoastronomy in the New World. Cambridge University Press, UK (1982)

Bacci, G.M., Coppolino, P.: La Necropoli di Abakainon, primi dati. EDAS, Messina (2009)

Bauer, B.S.: The Sacred Landscape of the Inca: The Cusco Ceque System. The University of Texas Press, Austin (1998)

Brea, B.L.: La Sicilia prima dei greci. Il Saggiatore, Milano (1958)

Brea, L.B.: Segnalazioni di rinvenimenti paleolitici in Sicilia. Bullettino di Paletnologia Italiana, 74, Roma, pp. 7–22 (1965)

Brea, L.B.: La necropoli di Longane. Bullettino di Paletnologia Italiana, 76, Roma, pp. 181–254 (1967)

Bianco, S.: Il culto delle acque nella preistoria. In: Nava, M.L. (ed.), Archeologia dell'Acqua in Basilicata, Soprintendenza Archeologica della Basilicata (1999)

Blaeu, G.: Sicilia Regnum. Amsterdam (1640)

Brun, J.-P.: Le vin et l'huile dans la Méditerranée antique: viticulture, oléiculture et procédés de transformation. Errance, Paris (2003a)

Brun, J.-P.: Archéologie du vin et de l'huile: de la préhistoire à l'époque hellénistique. Errance, Paris (2003b)

Bulifon, A.: Regno et Isola di Sicilia. (1692)

Campagna, B.: Recenti ricognizioni nel territorio di Rodì Milici. In: Calderone, A., Caltabiano, M., Fiorentini, G. (eds.), Archeologia del Mediterraneo. Studi in onorre di Ernesto De Miro, Roma (2003)

Cavalier, M.: La stazione preistorica di Tindari. Bullettino di Paletnologia Italiana, 79, Roma, pp. 61–94 (1970)

Cavalier, M.: Il riparo della Sperlinga di S. Basilio (Novara di Sicilia). Bullettino di Paletnologia Italiana, 80, Roma, pp. 7–76 (1971)

Cavalier, M.: Il riparo "Sperlinga" di S. Basilio. In: Martinelli, M.C., Spigo, U. (eds.), Tra i Peloritani e i Nebrodi prima dei Greci. Ricerche promosse e condotte da Luigi Bernabò Brea nei siti preistorici e protostorici della Sicilia nord-orientale, Lipari (2000)

Ciaceri, E.: Culti e miti nella storia dell'antica Sicilia. Arnaldo Forni, Sala Bolognese (BO) (2008)

Collignon, F.: Sicilia regnum. (1676)
Coronelli, V.M.: Trinacria oggidì Sicilia. Venezia (1696)
Curti, E., Mucciarelli, M., Polcaro, V.F., Prascina C., Witte, N.: The "Petre de la Mola" megalithic complex on the Monte Croccia (Basilicata). In: Rappenglück, M. (ed.), Proceedings of the SEAC 17th annual meeting, 25–31 Oct 2009, Alexandria, Egypt, B.A.R., London (2009)
Devins, P., Musco, A.: Argimusco Decoded, Lulu.com (2014)
De Wit, F.: Insula sive regnum siciliae. Amsterdam (1680)
Dufour, L., La Gumina, A.: Imago Siciliae. Domenico Sanfilippo Editore, Catania (1998)
Duval, P.: La Sicile, avec les anciens noms de presque tout ses places, rivieres, chasteaux, &c. et diverses observations nouvelles. Parigi (1672)
Fazello, T.: De Rebus Siculis decades duae, Palermo (1558)
Funcke, D.: Infelicis Regni Siciliae Tabula in Tres Valles Divisa Demonae, Notae et Mazarae. Norimberga (1693)
Ghezzi, I., Ruggles, C.: Chankillo, a 2.300 year-old Solar Observatory in coastal Peru, Science, 315, n. 5816, pp. 1239–1243 (2007)
Gori, D., Orlando, A.: The megalithic Temple of Diana on the Cefalù Rock (Sicily). Mediterr. Archaeol. Archaeometry **16**(4), 207–211. Napoli (2016)
Hannah, R., Magli, G., Orlando, A.: The role of urban topography in the orientation of Greek temples: the cases of Akragas and Selinunte. Mediterr. Archaeol. Archaeometry **16**(4), pp. 213–218. Napoli (2016a)
Hannah, R., Magli, G., Orlando, A.: Astronomy, topography and landscape at Akragas' Valley of the Temples. J. Cult. Herit. (2016b)
Hoskin, M.: Storia dell'Astronomia. BUR, Milano (2009)
Idrisi, N.: Il Libro di Ruggero. In: Rizzitano, U. (ed.), Palermo. Flaccovio Editore (2008)
ISPRA: Carta geologica d'Italia (2011)
Jaillot, A.H.: La Sicile divisée en trois Provinces ou Vallées sçavoir Valle di Demona, Valle di Noto, et Valle di Mazara. Parigi (1681)
Jansson, J.: Sicilia regnum (1640)
Lentini, M.C.: Naxos of Sicily: the First Colonial Settlement. In: "ASAA", vol. 14, pp. 455–470 (2006)
MacKie, E.: Maeshowe and the winter solstice: ceremonial aspects of the Orkney Grooved Ware culture. Antiquity **71**, 338–359 (1997)
Malaterra, G.: Imprese del conte Ruggero e del fratello Roberto il Guiscardo. In: Spinnato, E. (ed.), Palermo, Flaccovio Editore (2000)
Martinelli, M.C.: La necropoli di Longane: la tomba 21. In: Martinelli, M.C., Spigo, U. (eds.) Tra i Peloritani e i Nebrodi prima dei Greci. Ricerche promosse e condotte da Luigi Bernabò Brea nei siti preistorici e protostorici della Sicilia nord-orientale, Lipari (2000a)
Martinelli, M.C.: La stazione preistorica di Tindari. In: Martinelli, M.C., Spigo, U. (eds.) Tra i Peloritani e i Nebrodi prima dei Greci. Ricerche promosse e condotte da Luigi Bernabò Brea nei siti preistorici e protostorici della Sicilia nord-orientale, Lipari (2000b)
Martinelli, M.C.: Ucria: Riparo della Rocca S. Marco. In: Martinelli, M.C., Spigo, U. (eds.), Tra i Peloritani e i Nebrodi prima dei Greci. Ricerche promosse e condotte da Luigi Bernabò Brea nei siti preistorici e protostorici della Sicilia nord-orientale, Lipari (2000c)
Martinelli, M.C.: L'insediamento preistorico dell'età del Bronzo. In: Spigo, U. (ed.) Tindari, l'area archeologica e l'antiquarium. Milazzo, Rebus Edizioni (2005)
Martinelli, M.C., Prosdocimi, B.: Insediamento stagionale degli inizi dell'età del Bronzo sul Monte Bammina (in 630 s.l.m). In: Congiu, M., Miccichè, C., Modeo, S. (eds.), Eis acra: insediamenti d'altura in Sicilia dalla preistoria al III sec. a.C., Caltanissetta: Salvatore Sciascia Editore (2009)
Maurolico, F.: Descrizione dell'Isola di Sicilia, Venezia (1546)
McClusky, S.C.: Historical Archaeoastronomy: the Hopi Example. In: Aveni, A.F. (ed.) Archaeoastronomy in the New World. Cambridge University Press, UK (1982)
Melis, M.G.: Osservazioni sul ruolo dell'acqua nei rituali della Sardegna preistorica, pp. 111–124. Rivista di Scienze Preistoriche, LVIII (2008)

Mercator, G.: Siciliae Regnum. Bibliothèque nationale de France, département Cartes et plans, GE DD-2987 (5683) (1589)

Mercator, G.: Siciliae regnum. Amsterdam (1630)

Merian: Sicilia regnum. Francoforte (1635)

Metellus: Siciliae regnum (1601)

Nigro, L.: L'orientamento astrale del Tempio del Kothon di Mozia, in Atti del VII Convegno Nazionale della Società Italiana di Archeoastronomia, pp. 15–24 (2010)

Orlando, A.: Studio archeoastronomico della "spirale megalitica" di Balze Soprane di Bronte (CT): analisi preliminare. In: L'Acqua, la Roccia e l'Uomo, Puglisi, A., Turco, M. (eds.), Nicolosi (2015)

Orlando, A.: Archaeoastronomy in Sicily: Megaliths and Rocky Sites. In: Silva, F., Malville, K., Lomsdalen, T., Ventura, F. (eds.), The Materiality of the Sky. Sophia Centre Press (2016)

Orlando, A., Magro, M.T., Scaravilli, M.S.: The oriented altars of Rocca Pizzicata and the rocky sites of Alcantara Valley. Mediterr. Archaeol. Archaeometry **16**(4), 203–206. Napoli (2016)

Paladino G. (ed.): Bartholomaei de Neocastro Historia Sicula (1250–1293). In: Rerum Italicarum Scriptores, XIII, 3, Bologna (1921)

Pantano, G.M.: Megaliti di Sicilia. Edizioni Fotocolor, Patti (ME) (1994)

Pantano G.: Argimusco, la verità sull'origine del nome. Messina (2015)

Platania, V., Scaravilli, M.S.: Il complesso rupestre di Rocca Pizzicata. Agorà **45**, 123–125 (2013)

Portale, E.C.: Le nymphai e l'acqua in Sicilia: contesti rituali e morfologia dei votivi. In: A. Calderone (a cura di), Cultura e religione delle acque. Atti del Convegno Interdisciplinare "Qui fresca l'acqua mormora…." (S. Quasimodo, Sapph. fr. 2,5), Messina 29–30 marzo 2011, Roma, pp. 169–191 (2012)

Puglisi S.F.: La valle dei palmenti. Archeologia vitinicola e rupestre in Sicilia. Armando Siciliano Editore, Messina (2009)

Quad von Kinckelbach, M.: Sicilia. Ioan. Bussemecher excudit (1600)

Samperi, P.: *Iconologia della gloriosa Vergine madre di Dio Maria protettrice di Messina*, Messina (1644)

Saporetti C. and Varisco S., Il complesso megalitico di Porta Zilla. Geo-Archeologia (1999/2): pp. 35–43 (1999)

Sofia, G.: Abakainon. Nella dimora di Ade. La necropoli in contrada Cardusa a Tripi. Giambra Editori (2015)

Sofia, G.: Il processo di "romanizzazione" della Provincia di Sicilia, il caso del comprensorio tirrenico tra i torrenti Mazzarrà e Patrì-Termini. Giambra Editori, Terme Vigliatore (2016)

Spigo, U., Rizzo, C., D'Amico, E., Vanaria, M.G.: Francavilla di Sicilia. L'anonimo centro di età greca. L'area archeologica e l'antiquarium. Rubbettino Editore, Soveria Mannelli (CZ) (2008)

Stephen, A.M.: Hopi journal. In: Parsons, E.W.C. (ed.), Columbia University Press, New York (1936)

Terranova, N.: Montalbano Elicona, storia e attualità. Editer, Roma (1986)

Todaro, G.: Alla ricerca di Abaceno. Armando Siciliano Editore, Messina (1992)

Uggeri, G.: La Sicilia nella rotta tra Roma e l'Africa (III-IV sec. d.C.). In: Pagani e cristiani in Sicilia, Atti del X Congresso sulla Sicilia antica cristiana (Palermo-Siracusa 2001), Kokalos, XLVII-XLVIII, Roma, pp. 63–96 (2008)

Visscher, N.: Regnum siciliae. (1698)

Zuidema, R.T.: The Ceque System of Cuzco: he Social Organization of the Capital of the Inca. Brill, Leiden (1964)

Part II
Cultural Astronomy

The Palaeolithic Sky

Elio Antonello

Abstract Some results of sky simulations for the Palaeolithic epoch regarding stars and constellations are presented. A conservative estimate of the precision of the computed stellar coordinates is several arcminutes. Possible practical applications of such simulations should be seen in the context of collaborations among humanistic and scientific disciplines in the field of prehistoric archaeology.

1 Introduction

There is some interest by scholars in the simulation of the sky of tens of thousands of years ago, but in many cases the available software for representing the constellations give unreliable results for a time interval longer than some thousand years. The reason is the limited validity of the polynomial expressions in time usually adopted for computing the precession effect; as an example of unreliable results, Meeus (1998; p. 136) quoted the case of Polaris with a declination of −87° in the year 32700. There are available softwares that integrate directly the equations of dynamics and overcome the problem (e.g. SOLEX; Vitagliano 1996). However, we have tried to assess the issue by using the trigonometric formulas developed by Berger (1976, 1978), whose validity is of the order of some million years, even though those formulas were intended mainly for application to the study of palaeoclimate. We think that, as far as one is satisfied with a limited precision (several arcminutes) of the coordinates of the stars, this approach is sufficient.

We remark that, when talking about the relation between human beings and the sky of the far past, it would be important to take into account not only the results of archaeology and palaeoanthropology, but also the different environment and landscapes suggested by palaeoclimatology studies.

E. Antonello (✉)
Astronomical Observatory of Brera (INAF), Milan, Italy
e-mail: elio.antonello@brera.inaf.it

A. Orlando (ed.), *The Light, The Stones and The Sacred*,
Astrophysics and Space Science Proceedings 48,
DOI 10.1007/978-3-319-54487-8_9

2 Orbit

The eccentricity of the ecliptic, the obliquity or inclination of the Earth rotation axis with respect to the axis of the orbit, the intersection between the ecliptic and the equatorial plane, the longitude of the perihelion, and the inclination and longitude of the nodes of the ecliptic with respect to the solar system reference plane change with time. We used trigonometric expressions with amplitudes, frequencies and phase terms derived by Berger (1976) for those various parameters. The adopted epoch by that author was 1850.0, and we made a check of the results with those collected and published by him referred to 1950.0. Moreover, we made a comparison of the obliquity and general precession with those obtained with an accurate expansion in time (Laskar 1986), for a time interval of 10,000 years. We estimate an accuracy of some arcminutes, and the corresponding uncertainty in time of the given coordinates of a star in the sky is of several centuries. However, we remark that the evaluation of the precision of the results of the computations requires a comparison with real observations, and that means a time interval of just few centuries. Therefore, in principle the validity of the results very far in time cannot be assessed for any model of the Earth dynamics.[1]

We recall that, because of the change of obliquity, the true displacement of the mean north celestial pole during a precession cycle is not a circle; therefore, the figures illustrating the precession effect usually found on the web and in books could be misleading. Moreover, given the limited precision, it is not worth to try to get 'better' results by including other effects such as the nutation, since it amounts to less than ten arcseconds.

We remark also that, since a precession cycle does not repeat exactly, it is not possible to recover the same reference point in the sky after each cycle. One can estimate only a periodicity of the precession cycle, which will depend of course on the selected epoch and reference point. For example, if we consider the general precession and the epoch 1850.0, we get a periodicity of 25101 years for the previous cycle, and 26442 years for the current one.

3 Results

In a previous work (Antonello 2009) we have discussed the case of Ursa Major, Ursa Minor and Arcturus. Here we will show some other examples of the simulations for the year 54000 BC. The sky seen from the latitude of +45° is shown in Fig. 1; given the high proper motion, Arcturus was closer to the stars Vega, Altair

[1]We were recently informed about the paper by Vondrak et al. (2011), where both trigonometric and polynomial expressions have been used.

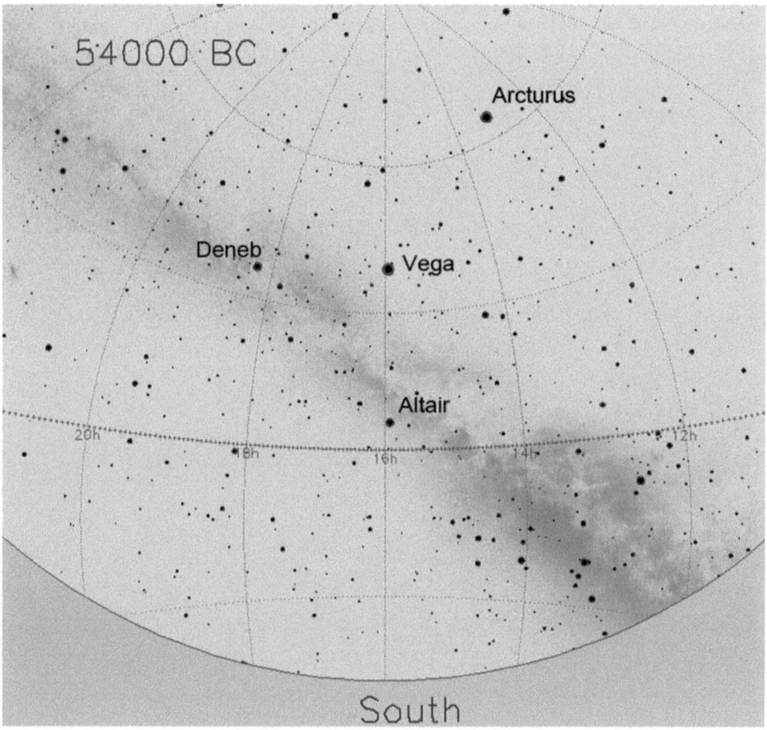

Fig. 1 The sky with the *bright* stars Arcturus, Vega, Deneb and Altair seen from the northern latitude of 45°; the epoch is 54000 BC

and Deneb than today. In Fig. 2, Sirius appears on the background of the Milky Way; we have marked Capella (Auriga constellation), and one can note that the nearby Gemini constellation includes Procyon. The southern sky seen from the latitude of +10° is shown in Fig. 3; the Crux has a different shape from today, while Alpha Centauri is far from Beta Centauri and fainter than today, given its larger distance.

4 Homo Sapiens and Palaeoclimate

During the last decades impressive progresses have been made in the genetic studies as regards the evolution of our species. The first modern humans should have appeared in Africa about 200,000 years ago (McDougall et al. 2005), and during the past 100,000 years our ancestors interbred with Neanderthals, out of Africa, as shown by the genome of living Asians and Europeans (Gibbons 2011). Recent data suggest that *Homo sapiens* was in Western Siberia about 45,000 years

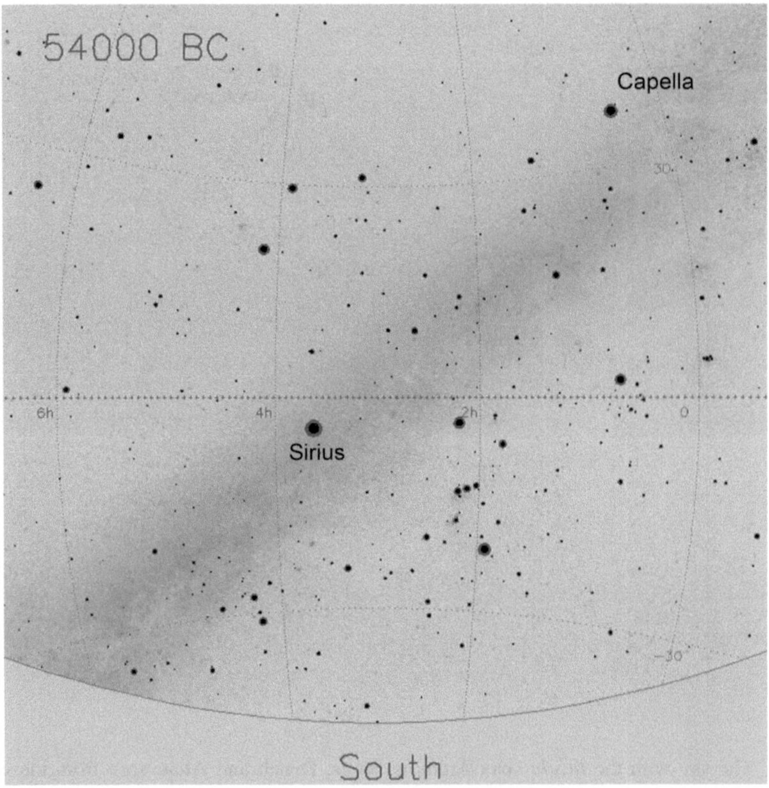

Fig. 2 Sirius, Capella and some nearby constellations (Auriga, Gemini, Taurus and Orion) seen from the northern latitude of 45°. Procyon is close to Gemini (Castor and Pollux, the *bright* stars above Sirius)

ago, and the admixture between his ancestors and Neanderthals should have occurred approximately 50,000 to 60,000 BP (Fu et al. 2014).

When studying the possible relation between our ancestors and the sky, we should take into account the very different environment of the Palaeolithic, given the different climate and length of the seasons. The length of the astronomical seasons depends on the eccentricity of the Earth orbit: a larger eccentricity means a larger difference of the season length; moreover, the seasons 'rotate' on the ecliptic plane with a period of about 19,000–23,000 years, an effect named climatic precession. During the Palaeolithic epoch, for several tens of thousands of years, the mean temperature was lower than today, and this was, at different level, a global climatic trend. During the last glacial maximum, the European landscape was mainly covered by a cool steppe vegetation, with some forest stands in Mediterranean regions and tundra shrubs in France (Peyron et al. 1998). The lower temperature implies an anomaly, that is, a difference from the present day average temperature, probably much larger during winter (Davis et al. 2003).

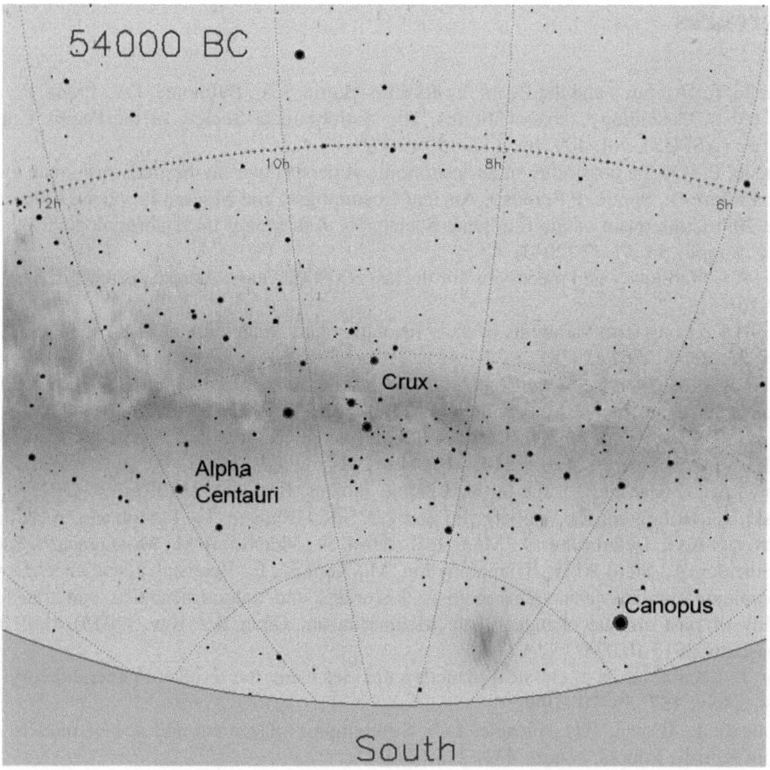

Fig. 3 Southern constellations seen from the latitude +10°. Crux has a different shape from today, while Alpha Cen is very far from Beta Cen

A new approach is probably needed for a better understanding of the impact of climate changes on the evolution of our species and of the human civilization, in a way similar to that proposed recently by Izdebski et al. (2015) for the case of the Mediterranean in the past millennia. The authors discussed at length the theoretical and practical issues involved in the collaboration among the natural sciences and the humanities, or social sciences, focused on the study of human perceptions and actions in relation to climate and environmental change. In particular, they analysed the contribution of archaeology (humanities/social sciences), history (humanities) and palaeoenvironmental sciences (natural sciences). They did not mention astronomy, or archaeoastronomy. However, it is well known that the long-term climate changes are triggered by the astronomical parameters (orbital forcing), and apparently there is no plausible alternative to that theory. An astronomical approach could be probably of help in this context (Antonello 2013), even though it seems that the way for an effective collaboration among those disciplines is quite long.

References

Antonello, E.: Arcturus and the Bears. In: Rubiño-Martin, J.A., Belmonte, J.A., Prada, F., Alberdi, A. (eds.), Cosmology across Cultures, The Astronomical Society of the Pacific Conference Series (ASPCS), vol. 409, pp. 157–159 (2009)

Antonello, E.: On the possibility of an 'astronomical perspective' in the study of human evolution. In: Culture, I., Šprajc, P.P. (eds.), Ancient Cosmologies and Modern Prophets, Proceedings of the 20th Conference of the European Society for Astronomy in, Anthropological Notebooks, XIX Suppl., pp. 41-47 (2013)

Berger, A.L.: Obliquity and precession for the last 5000000 years. Astron. Astrophys. **51**, 127–135 (1976)

Berger, A.L.: Long term variations of daily insolation and Quaternary climatic changes. J. Atmos. Sci. **35**, 2362–2367 (1978)

Davis, B.A.S., Brewer, S., Stevenson, A.C., Guiot, J., et al.: The temperature of Europe during the Holocene reconstructed from pollen data. Quat. Sci. Rev. **22**, 1701 (2003)

Fu, Q., Li, H., Moorjani, P., Jay, F., et al.: Genome sequence of a 45,000-year-old modern human from western Siberia. Nature **514**, 445–449 (2014)

Gibbons, A.: A new view of the birth of Homo sapiens. Science **331**, 392–394 (2011)

Izdebski, A., Holmgren, K., Weiberg, E., Stocker, S.R., Büntgen, U., Florenzano, A., Gogou, A., Leroy, S.A.G., Luterbacher, J., Martrat, B., Masi, A., Mercuri, A.M., Montagna, P., Sadori, L., Schneider, A., Sicre, M.A., Triantaphyllou, M., Xoplaki, E.: Realising consilience: How better communication between archaeologists, historians and natural scientists can transform the study of past climate change in the Mediterranean. Quat. Sci. Rev. (2015). doi:10.1016/j.quascirev.2015.10.038

Laskar, J.: Secular terms of classical planetary theories using the results of general theory. Astron. Astrophys. **157**, 59–70 (1986)

McDougall, I., Brown, F.H., Fleagle, J.G.: Stratigraphic placement and age of modern humans from Kibish, Etiopia. Nature **433**, 733 (2005)

Meeus, J.: Astronomical Algorithms, 2nd edn. Willmann-Bell Inc, Richmond (1998)

Peyron, O., Guiot, J., Cheddadi, R., Tarasov, P., et al.: Climatic reconstruction in Europe for 18,000 yr B.P. from pollen data. Quat. Res. **49**, 183 (1998)

Vitagliano, A.: Numerical integration for the real time production of fundamental ephemerides over a wide time span. Celest. Mech. Dyn. Astron. **66**, 293–308 (1996)

Vondrák, J., Capitaine, N., Wallace, P.: New precession expressions, valid for long time intervals. Astron. Astrophys. **534**(A22), 1–19 (2011)

Astronomy in the Odyssey: The *Status Quaestionis*

Salvo L. Guglielmino, Paolo B. Cipolla and Innocenza Rizzo Giudice

Abstract Homeric poems, other than being the first literary record of the ancient Greek civilization, represent a rich source of information about the "scientific" knowledge of Greeks during the Archaic period. In the *Iliad* and in the *Odyssey*, whose redaction dates back to the VIII century BC, there are mentions of several heavenly bodies: the Sun and the Moon, the planet Venus, some stars, constellations (Boötes, Ursa Major and Orion) and asterisms (Pleiades and Hyades). Hence, the Homeric epos is an important evidence for archaeoastronomical studies dealing with civilizations that arose in the Mediterranean basin, in particular ancient Greece. However, few scholars have addressed so far a systematic review of the Homeric passages that include astronomical references, on the one hand because of little interest in such kind of problems in philological studies, on the other hand since astronomical expertise is required to support the analysis. By virtue of the subject matter of the story itself—the nostos of Odysseus back to Ithaca—the *Odyssey* contains the majority of the passages concerning the description of celestial bodies and phenomena. This supports indications, obtained independently, that during the pre-Homeric period stars had already been extensively used as a reference for navigation, in order to find and follow the route during the long crossings of the Mediterranean sea. In the present contribution we describe the state of the art of scientific and philological research concerning the passages of the *Odyssey* where astronomical knowledge of that time is reflected. Some of them have been interpreted as the description of celestial events that really occurred. This has given rise to a lively discussion among the scientific community on the possibility to assign a date to the historical facts

S.L. Guglielmino(✉)
University of Catania – Dipartimento di Fisica e Astronomia,
Sez. Astrofisica, Via S. Sofia 78, 95123 Catania, Italy
e-mail: salvo.guglielmino@oact.inaf.it; sgu@oact.inaf.it

P.B. Cipolla · I. Rizzo Giudice
University of Catania – Dipartimento di Scienze Umanistiche,
Piazza Dante 32, 95124 Catania, Italy
e-mail: pcipolla@unict.it

A. Orlando (ed.), *The Light, The Stones and The Sacred*,
Astrophysics and Space Science Proceedings 48,
DOI 10.1007/978-3-319-54487-8_10

that were recounted by Homer, in the first place the Trojan War, based on the astronomical phenomena that were observed in the Mediterranean at the end of the II millennium BC. Finally, we discuss new prospects for archaeoastronomical research on the Homeric epic.

1 Introduction

In its early stage of development, archaeoastronomy presented a dichotomy between the investigations on observable alignments, astronomically oriented, at prehistoric sites in the Old World and the analyses of time-reckoning carried out on the basis of written or painted records in the New World (Aveni 1989).

During the second half of the 20th century, the New World researchers developed an anthropology of astronomy, revealing complex connections between astronomical practices and politics, economics, and social organization of Meso- and South American cultures. At the same time, some archaeologists in Europe investigated the alignments in standing stones at megalithic sites, by following the experiences of Thom and Hawkins in Britain. Their main aim was to determine whether celestial events at the horizon might have influenced building construction and to seek out astronomical orientations that fit the alignments.

Classical Old World scholars adopted a condescending attitude towards these investigations, sometimes even regarded with suspicion. By contrast, they confined themselves to the decipherment and analysis of literary documents, pursuing the attitude of the textual critics of the 19th century, and elaborating a history of astronomy. This was reflected in the ample literature concerning the progress of western astronomy from a pre-scientific to a scientific stage (Heath 1932; Dicks 1970; Aaboe 1974; Parker 1974; Neugebauer 1975; Price 1975). However, its development was removed from the historical context and considered only in terms of the degree of precision achieved in observations and predictions (Ruggles 1984; Aveni 1989).

This separation led to a striking paradox. Although the text-rich civilizations of the Babylonians, Egyptians, and Graeco-Romans offered an overwhelming amount of astronomical information in their records, only few studies investigated on the possible presence of astronomical references in the architecture and iconography of these cultures and in their literature not explicitly dealing with celestial phenomena.

Nowadays the situation has changed. It has become clear that archaeoastronomy is an interdisciplinary field of research that aims to investigate the relationship between astronomical phenomena and cultural behaviour, studying the practice and use of astronomy among the ancient cultures on the basis of all forms of *written* and *unwritten* evidence (Aveni 1987, 1989). This approach requires a combination of methods deriving from diverse disciplines, integrating astronomy with archaeology, cultural anthropology, and linguistics, as well as historical sciences (Ruggles and Urton 2006), being ultimately part of the wider discipline of cultural astronomy.

However, this new research paradigm in archaeoastronomical investigations struggles to make inroads in studying ancient Greek culture. The long-standing division between the classicists' study of astronomy as a science, belonging to the domain of the literate historians, and the archaeologists' focus on material records, like alignments or artefacts, is still deep. This is pointed out, for instance, in the scant number of articles concerning archaeoastronomy in ancient Greece with respect to other cultures, that are present in the recent masterful essay by Ruggles (2015).

Aveni (1987) highlighted the peculiarity of the ancient Greek civilization during the initial discussions about the discipline in Italy. The focus of modern scholars on the Eudoxus' surviving fragments or the *Almagest* e.g. (Heath 1932; Neugebauer 1975; Evans 1998; Aaboe 2001; North 2008) is son of the separation between a highly intellectualized view of the cosmos, which was developed in the 5th century BC by Platonic philosophy and eventually gave rise to Ptolemaic cosmology, and peasant astronomy. In fact, some texts of the Archaic period suggest that naked-eye observations of the sky played an important role in the everyday life, using the cycles of celestial bodies to regulate seasonal activities. For the sake of example, Hesiod's *Works and Days* includes an agricultural calendar (vv. 383–617), demonstrating that Greeks in the 8th century BC used the annual appearance of stellar phenomena to identify the appropriate times for ploughing and harvesting. The long mention of this practice in a poem addressed to a wide audience makes manifest that the scene was very familiar to the public. But then, after the progress of philosophical speculation, the astronomical knowledge of earlier ages, based on the direct observation of celestial phenomena, was relegated to a primitive past.

The interest of most scholars who studied Archaic Greek astronomy, usually philologists, has been limited in obtaining information about the temporal setting of the narrative from astronomical references found in the texts. Instead, Aveni (1987) recommended to examine in more detail the sources, benefiting from Archaic literature still embedded in the oral tradition (Havelock 1963), to carry out a new analysis and to formulate hypotheses regarding the possible astronomical orientation of certain alignments in buildings of the Hellenic and pre-Hellenic periods.

Actually, early investigations of temples orientations, carried out in the 19 and 20th centuries, spread out the idea that ancient Greeks usually built their sacred monuments facing the east towards the rising sun.[1] However, some interpretative reconstructions were often detached from the archaeological evidence and hence were harshly criticised e.g., (Herbert 1984). Thus, the issue concerning the orientation of ancient Greek temples is still under debate (Aveni and Romano 2000; Salt 2009), while the analysis has also extended to civic buildings, such as theatres.[2]

At the moment, the current status of the research on ancient Greece is slowly evolving, as some scholars with different expertise have finally begun to collaborate in the field. In this context, it is worth noting that Boutsikas and Ruggles (2011) have recently shown the potential for archaeoastronomy in Archaic Greece by investigating the astronomy in Alcman's Partheneion, written in the 7th century BC, and its possible relationship with the rituals performed in honour of Artemis Orthia at her sanctuary in Sparta. Adopting an approach which integrates and contextual-

[1]For more details, see, e.g., the contribution of Hannah et al. (2017).

[2]See, e.g., the contribution of Carnevale et al. (2017).

izes archaeological and literary evidence with astronomical data, they have obtained promising results on understanding the role of astronomy in ancient Greek religious practice and perception of the cosmos.

2 Astronomy in the Homeric Poems

A fairly good number of astronomical references appear in the Homeric poems[3]: they include the sky, the Sun and the Moon, the planet Venus, stars, such as Sirius, some constellations (Orion, Boötes, Ursa Major) and asterisms (star clusters like the Pleiades and the Hyades). These astronomical allusions were known since antiquity: the pre-Socratic philosopher Heraclitus mentioned Homer as ≪the Astronomer≫.[4]

The ancient exegetical tradition attributed all knowledge (*polymatheia*) to Homer: the poet was considered to be the father of all the sciences.[5] Scholia and allegorical commentaries often reflect interest about astronomy: an example is given by the pseudo-Plutarch text *De Homero*, which likely dates back to the 2nd century AD.[6]

---

[3] All Greek quotations are based on the most recent critical edition of the texts: van Thiel, H., ed.: Homeri Odyssea/Homeri Ilias. (Hildesheim, Zürich, and New York, 1991/1996).

[4] ∑ *Il.* 18, 251 p. 481 Erbse (= Heraclit. *Frag.* 105 Diels–Kranz)

> (*251b.*) ***AT*** Ἡράκλειτος ἐντεῦθεν ἀστρολόγον φησὶ τὸν Ὅμηρον κτλ.

[5] Cf. Str. 3, 4, 4. For more details about *polymatheia*, see, e.g., Broggiato (2001), 103–105.

[6] Cf. [Plu.] *Vit. Hom.* 2, 104–108. At 106–107, for instance, we read about constellations:

> (*106*) κτλ. καὶ τὴν Ἄρκτον τὴν ἀεὶ στρεφομένην περὶ τὸν ἀειφανῆ πόλον τὸν βόρειον καὶ διὰ τὸ μετέωρον μὴ ἁπτομένην τοῦ ὁρίζοντος, ὅτι ἐν ἴσῳ χρόνῳ ὅ τε μικρότατος κύκλος, ἐν ᾧ ἐστιν ⟨ἡ⟩ Ἄρκτος, καὶ ὁ μέγιστος, ἐν ᾧ ⟨ὁ⟩ Ὠρίων στρέφεται ἐν τῇ τοῦ κόσμου περιφορᾷ, καὶ τὸν βραδέως δυόμενον Βοώτην, ὅτι πολυχρόνιον ποιεῖται τὴν κατάδυσιν, οὕτω πεπτωκὼς θέσει, ὥστε ὀρθὸν καταφέρεσθαι καὶ συγκαταδύεσθαι τέσσαρσι ζῳδίοις, τῶν πάντων εἰς ὅλην τὴν νύκτα μεριζομένων ἓξ ζῳδίων. εἰ δὲ μὴ πάντα τὰ περὶ τῶν ἄστρων θεωρούμενα διεξῆλθεν, ὡς Ἄρατος ἢ ἄλλος τις, οὐ χρὴ θαυμάζειν· οὐ γὰρ τοῦτο προέκειτο αὐτῷ. (*107*) Οὐκ ἀγνοεῖ δὲ τὰς αἰτίας τῶν περὶ τὰ στοιχεῖα συμπτωμάτων, οἷον σεισμῶν καὶ ἐκλείψεων. κτλ.

> [106] [...] [He also knew] the Bear, always revolving around the permanently visible north pole star and never passing below the horizon, because of its elevation. He knew that the smallest circle (that of the Bear) and the largest (that of Orion) turn in the revolution of the cosmos in exactly the same time. He knew that Boötes is slow to set, since he makes its setting take a long time, because its position is suche that it is borne along upright to set with four zodiacal signs while the entire night is divided among six signs. If he did not go into the entire matter of astronomy as for instances Aratus does, this is no wonder, since that is not his subject. [107] He is not ignorant of the causes of such events affecting the elements as earthquakes and eclipses. [...]

The translation is taken from Keaney, J.J., Lamberton, R.: Plutarch. Essay on the Life and Poetry of Homer (Atlanta 1996).

In modern times, quite some studies investigated astronomical knowledge in Homer's *Iliad* and *Odyssey* (see, e.g., Theodossiou et al. 2011, and references therein). Astronomers have recently developed a new interest in the field, especially when simulations of the sky at different times have been made possible by computation (e.g., Baikouzis and Magnasco 2008; Papamarinopoulos et al. 2012, 2013). Philologists, for their part, have rediscovered this exegetical line bringing new arguments for the analysis of certain passages (e.g., Broggiato 2003; Renaud 2003; Finkelberg 2004; Dettori 2008; Catalin 2010; Gainsford 2012; Revello 2013).

Despite these contributions, a systematic review of the topic from an archaeoastronomical point of view is still lacking. One of the main issues is that most studies are conducted from a single perspective, either literary or scientific. This fact leads frequently to misinterpretations, when the text is analysed using approaches in which some important aspects of the other discipline are neglected or ignored.

To make the situation worse, other authors proposed that the Homeric poems are, in reality, fundamentally astronomical texts. Characters and events described in the *Iliad* and *Odyssey* would be astronomical metaphors. For instance, Odysseus would be a solar hero, returning to Penelope as soon as the nineteenth year was complete, at the end of a metonic cycle (Murray 1924). His twelve adventures would have deep connections with the twelve signs of the zodiac (Grey 2013). Numerical data in the narrative would contain accurate knowledge of astronomical cycles and calendars (Wood and Wood 2011). The journey of the hero could be also related to the twenty-year period between two successive conjunctions of Jupiter and Saturn (Gendler 1984), or to the annual period between the heliacal setting and rising of Sirius (Ropars 2003). Most of these works seem, nevertheless, rather speculative and underwent serious criticism for not being based on coherent supporting claims.

Apart from these interpretations, the *Iliad* and *Odyssey* give the opportunity to study the importance of astronomical phenomena in pre-Homeric and Homeric times. The oral-formulaic theory developed by Parry and Lord showed that the Homeric poems represent the culmination of a long oral tradition. For this reason, the analysis can also benefit from the contributions of linguistics and anthropology, as suggested by Aveni (1987).

An extensive survey of the passages with astronomical information was made by Dicks (1970), who started his treatise on early Greek astronomy from Homer. Given his focus on the development of astronomy as a science, he affirmed that celestial phenomena were well known as purely observed recurrences, but that they were not understood in terms of a rational explanation.

The most extended passages with astronomical content in the *Iliad* and *Odyssey* are the description of the Achilles' shield (*Il.* 18, 483–489[7]) and the Odysseus' departure from Ogygia (*Od.* 5, 270–277[8]), respectively. Various stellar groups are named: Orion, the two bright star clusters of the Pleiades and Hyades, Boötes, and the (Great) Bear, also called the Waggon. The mention of these constellations seems to be related to human activities like agriculture and navigation (see Sect. 3.1).

[7]Cf. Edwards (1991) ad loc.; see also (Phillips 1980; Hannah 1994).

[8]Cf. Heubeck et al. (1988) ad loc.; see also Hannah (1997).

Stars fill up the heavens: the formula οὐρανὸς ἀστερόεις, 'starry sky', occurs 11 times,[9] while the adjective ἀστερόεις is used in a different context only two more times. The brightest stars appear in similes for gods and men, so they were sufficiently familiar to the audience.

References[10] to the περιτελλόμενον ἔτος, 'revolving year', and to the recurring seasons (ὧραι) suggest that the Homeric calendar was already based on the solar (seasonal) year (Fisher 1977). There is debate whether these mentions reflect knowledge of solstices: in actuality, «τροπαὶ ἠελίοιο», 'turnings of the sun', appear in *Od.* 15, 404, but the interpretation is rather controversial.[11] However, the verb τρέπω, 'to revolve', which is used for recurring seasons, is a derivative form the same root as the word 'turnings', **tr̥p-/trep-/trop-*.

As a further argument, we note in the Homeric poems the usage of two competing words, ἐνιαυτὸς and ἔτος, with the same meaning, 'year' (Emlyn-Jones 1967). Beekes (1969) observed that it is possible there was an original difference which later disappeared: in particular, ἐνιαυτὸς has a root connected with ἰαύω, 'to rest', so that the sense is closer to 'anniversary', 'cycle between days'.[12] The affinity with the Latin word *solstitium* is fairly evident. Actually, the dramatic action of the *Odyssey* begins with a temporal notation «ἀλλ' ὅτε δὴ ἔτος ἦλθε περιπλομένων ἐνιαυτῶν» (*Od.* 1, 16), which forces us to emphasise the semantic difference between the two words in the translation: 'but as years passed away (lit. turned again), when the *very* year came', a certain year in which gods settled that the hero should return home. Homer's listeners would have noticed the different connotations of the words.[13]

This brief review clearly shows that, beyond the criticism of some scholars (Dicks 1970; Fisher 1977), allusions to celestial events suggest a comprehension of the astronomical phenomena at a level deeper than mere observation. Furthermore, astronomers should be aware that an archaeoastronomical analysis of these complex texts requires a sound philological approach to avoid misunderstandings.

2.1 *Orion and Sirius*

An illuminating example of a fruitful application of linguistics to problems of archaeoastronomy in Homer is offered by the verses referring to Orion. In antiquity, he has been mentioned both as a hero of ancient times and as a constellation. In the

[9]Cf. οὐρανοῦ ἀστερόεντος *Il.* 5, 769; 6, 108; 8, 46; 19, 130; *Od.* 20, 113; -ῷ -εντι *Il.* 4, 44; -ὸν -εντα *Il.* 15, 371; 19, 128; *Od.* 9, 527; 11, 17; 12, 380.

[10]Cf. *Od.* 11, 294–295 = *Od.* 14, 293–294.

[11]Cf. Heubeck and Hoekstra (1989), ad loc.

[12]Cf. Chantraine (1968), s.v.

[13]Austin (1975), 240; see also Heubeck et al. (1988) ad loc.

Homeric poems, Orion is the only hero who is referred to in such a way,[14] although the poet never alludes to his transformation into a constellation.

In the Greek myth, Orion was a gigantic hunter. No ancient source tells his entire story, and various episodes differ from version to version.[15] According to the most common one, Orion was native to Boeotia and was son of Poseidon, who gave him the power of walking on the sea surface. He visited Chios where he fell in love with Merope and was blinded by her father, Dionysus's son Oenopion. He regained his sight by walking toward the rising sun to the East, and later lived in Crete, where he took up hunting. There Orion fell in love with the goddess Artemis, accompanying her through the woods, but was eventually in conflict with her, as he declared his willingness to slay every wild beast in the island.[16] Then, he was killed by an arrow of Artemis or by a sting of a giant scorpion. Upon his death, Zeus raised Orion to the stars and transformed him into a constellation, and the Scorpion as well.

The origin of Orion's name is obscure. In a later version of the myth, told by Ov. *Fast.* 5, 493–536, we have a hint about the derivation, since the name appears to be derived from the unusual manner in which he was conceived. Zeus, Poseidon and Hermes rewarded the elderly Orion's father, Hyrieus, for having given hospitality to them and poured their semen into the skin of a bull, from which the child was born without a mother. Hyrieus named the infant Ourion (cf. οὐρέω, 'to urinate'), which in time became Orion.

However, in several sources the hero's name, which is spelled Ὠρίων with ῑ in the Homeric epic, sometimes is attested as Ὠαρίων, with ᾰ and ῐ. In particular, this form appears in Eur. *Hec.* 1103,[17] and the scholia inform us that it results from a pleonastic addition.[18]

Pasquali (1934) already noticed this problem and recommended to reconstruct the reading Ὠᾰρῐ- whenever it occurs in the Homeric poems.[19] In fact, in Homer and Hesiod the ῑ always appears in an unstressed syllable of the foot. It is hence possible to substitute Ὠρῑ́ων into Ὠᾰρῐ́ων, with ᾰ and ῐ, at every occurrence.

For the sake of clarity, the scansion of *Il.* 18, 486[20] is given

Πληϊάδας ϑ' Ὑάδας τε τό τε σϑένος Ὠρίωνος
—∪∪|—∪∪|— |∪∪|—∪∪|——|—⌒

[14]Ὠρίων appears as a hero in *Od.* 5, 121; 11, 310; 11, 572; as a constellation in *Il.* 18, 486; 18, 488 = *Od.* 5, 274. More ambiguous is the mention in *Il.* 22, 29 (see below).

[15]Cf. *Od.* 11, 310; 11, 572-575; Eratosth. *Cat.* 32; Arat. *Phaen.* 636-646; Apollod. 1, 25; Ov. *Fast.* 5, 493–536.

[16]Note that an Artemis type deity has been correlated with the worship of πότνια ϑερῶν, 'the Mistress of the Animals', depicted in Minoan art (Chadwick 1976, 92; Cline 2010, 151–153; 267; 271–272); cf. *Il.* 21, 470.

[17]The form is also found in Pind. *N.* 2, 12; Call. *Dian.* 265; Nic. *Th.* 15; and in some fragments of Pindar, Corinna, and Callimachus; cf. Heubeck et al. (1988) on *Od.* 5, 274.

[18]∑ Eur. *Hec.* 1103 p. 82, 7 Schwartz: Ὠρίων καὶ κατὰ πλεονασμὸν τοῦ ᾱ Ὠαρίων Note that the ᾱ symbol, that is used in the scholia, indicates the α as a letter and does not refers to the vowel quantity.

[19]See p. 246 in the 1988 edition of the text.

[20]Hes. *Op.* 615 is almost parallel to *Il.* 18, 486, suggesting a formulaic origin for the verse.

Due to the presence of Ὠρΐωνος, the line is a spondaic hexameter, having a spondee instead of a dactyl in the fifth foot, that is infrequent in Homer.[21]

Concerning the constellation, we know that its heliacal rising occurred close to summer solstice at the end of the 2nd millennium BC. For ancient Greeks and other Mediterranean peoples, it marked the beginning of a period which is favourable for reaping and harvesting, but which is also characterized by intense heat that often leads to fatigue, fever, and illness. For this reason, the rise of Orion took on an ambivalent meaning for ancient Mediterranean civilizations.

Renaud (2003) pointed out that the constellation and the mythical hero share many traits, as both have a dual significance related to flourishing, but to excessive strength as well. In effect, Orion is somehow representative of a new humanity, as a hunter and civilizing hero, but he falls because of his pretentious attitude towards wild nature. Considering the etymology, he revealed that the ancient myth is probably hidden in Orion's name, suggesting that the latter is closely connected with this constellation characteristic of the summer season (see also Renaud 2004a, b).

First, he analysed the derivation of the hero's name Renaud (1996). The original form Ὠαρίων is a derivative of the root *ωαρ, outcome of the Indo-European root **ōsar* that means 'summer'. In particular, the form shows a suffix -ῐων, deriving from the intensifying suffix *-iōs*, that is also involved in the formation of the comparative. The vowel lengthening of the ι might have occurred in analogy to the existence of Greek anthroponyms ending in -ῑων, deriving from -ῐϝων. Therefore, the meaning of the word Ὠαρίων should be 'the Man of Summer'.

The heliacal rising of Orion, as stated above, coincided with the beginning of the summer. The proposed etymology might then show that Ὠαρίων referred to the constellation, before being the name of a mythical character. With the passing of time, Greek mythology might have created a hero, who was ambiguous in his deeds and characterized by excesses, bearing the same significance as the constellation.

Thus, Renaud (2003) reached the conclusion that Homer was acquainted with the story of Orion and his placement in the heavens, since he knew the double function of Orion for ancient Greeks, although he did not explicitly refer to it in his poems.

This interpretation is further supported by the fact that Sirius, which is mentioned as Orion's dog (*Il.* 22, 29), is characterized by being the star «ὅς ῥά τ' ὀπώρης εἶσιν», 'that rises in the late summer' (*Il.* 22, 27). This same concept is signified by ἀστὴρ ὀπωρινός (*Il.* 5, 5). The root *ωαρ is easily recognisable in these expressions.

Indeed, the heliacal rising of Sirius occurred about 50 days after the summer solstice, during the hottest part of the season. Late summer and early autumn were unhealthy periods because of malarial fever in the Mediterranean region. This explains why the star is said to be «οὔλιος», 'baleful' (*Il.* 11, 62; cf. ὄλλυμι, 'to lead to ruin'), as it shines ominous bringing fever to wretched mortals (*Il.* 22, 30–31).

[21] A similar consideration was made by Chantraine (1942), 16: the form κύν' Ὠρΐωνος which is found in *Il.* 22, 29 replaced the original κύν' Ὠᾰρΐωνος; cf. Richardson (1993) ad loc.

3 Astronomy in the *Odyssey*

The narrative of Odysseus' nostos and his seafaring adventures includes, overtly or more subtly, sundry references to celestial bodies and phenomena (e.g., the Moon, Venus, solstices). In the following, we will briefly illustrate two case studies.

3.1 The Departure of Odysseus from Ogygia

In book V, Hermes is sent to Calypso's island to tell her that Odysseus must leave. Then, Odysseus sails away from Ogygia, after having received detailed instructions from the nymph (*Od.* 5, 270-277):

αὐτὰρ ὃ πηδαλίῳ ἰθύνετο τεχνηέντως
ἥμενος· οὐδέ οἱ ὕπνος ἐπὶ βλεφάροισιν ἔπιπτε
Πληϊάδας τ' ἐσορῶντι καὶ ὀψὲ δύοντα Βοώτην
Ἄρκτον θ', ἣν καὶ ἅμαξαν ἐπίκλησιν καλέουσιν,
ἥ τ' αὐτοῦ στρέφεται καί τ' Ὠρίωνα δοκεύει,
οἴη δ' ἄμμορός ἐστι λοετρῶν Ὠκεανοῖο·
τὴν γὰρ δή μιν ἄνωγε Καλυψώ, δῖα θεάων,
ποντοπορευέμεναι ἐπ' ἀριστερὰ χειρὸς ἔχοντα.

and taking his seat artfully with the steering oar he held her
on her course, nor did sleep ever descend on his eyelids
as he kept his eye on the Pleiades and late-setting Boötes,
and the Bear, to whom men give also the name of the Wagon,
who turns about in a fixed place and looks at Orion,
and she alone is never plunged in the wash of the Ocean.
For so Kalypso, bright among goddesses, has told him
to make his way over the sea, keeping the Bear on his left hand.[22]

These lines mention some of the most spectacular constellations visible in the sky above the Mediterranean sea: the star cluster of the Pleiades, Boötes, Orion and the Bear, Ἄρκτον. The latter is the only one (οἴη) that does not bathe in Ocean: this indicates that it was known as circumpolar. Moreover, the simultaneous presence of the Pleiades and of the «ὀψὲ δύοντα», 'late-setting', Boötes suggests that the action is taking place during the spring or, more likely, during the autumn season.[23]

This sole passage in the Homeric poems contains a reference to stellar navigation. Keeping the northern constellation of the Bear on his left indicates that Odysseus should sail eastwards.[24] Indeed, Ursa Major was the traditional guide for Greek navigators: Homeric formula ἑλίκωπες Ἀχαιοί[25] was interpreted as "those who look at Helice", another name of the Great Bear related to the constellation's rotation.

[22] The translation is taken from Lattimore, R.: The Odyssey of Homer (New York, 1965).

[23] Cf. Heubeck et al. (1988) ad loc.; Austin (1975), 240–244, Catalin (2010).

[24] Cf. Heubeck et al. (1988) ad loc.; Hannah (1997).

[25] Cf. ἑλίκωπες Ἀχαιοί *Il.* 1, 389; 3, 190; 24, 402; -ας -ούς *Il.* 3, 234; 16, 569; 17, 274.

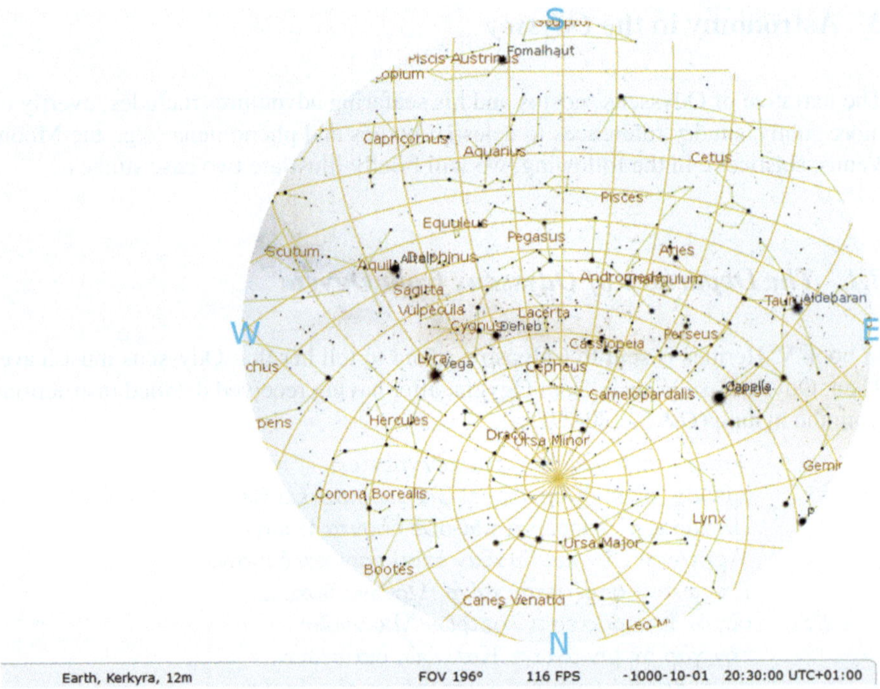

Fig. 1 Stellar map showing the position of the stars observed from Kerkyra (Corfu) island at 8:30 pm (7:30 pm local time) on 1 October 1000 BC. The view is toward the north. It is clearly visible that the angular distance of Ursa Major and Minor from the north celestial pole was about the same. Map created using Stellarium software v0.14.2

Nevertheless, by analysing ancient scholia, Finkelberg (2004) called into question the identification of the Bear. According to her, on the basis of a passage of Aratus[26] the expression «ἥ τ' αὐτοῦ στρέφεται», 'she turns about in the same spot' (*Od.* 5, 274) hints at Ursa Minor, turning around the pole; conversely, Ursa Major's orbit ran around the polar circle. Thus, she deduced that the Homeric text was tampered in the early Alexandrian period, in order to credit the poet with knowledge of Ursa Minor, which had been introduced to the ancient Greeks by Thales of Miletus.

Catalin (2010) objected to this conclusion, by considering the second part of the line where is said that the Bear «Ὠρίωνα δοκεύει», 'watches for Orion'. According to the theriomorphic representation of the constellations, the bodies of the Bears lie in opposite directions.[27] Ursa Minor looks towards the direction pointed by the Ursa Major's tail, in front of Boötes. Being Orion opposed to Boötes in the sky, the Bear 'watching for Orion' is Ursa Major.

[26]Cf. Arat. *Phaen.* 36–44.

[27]Cf. Arat. *Phaen.* 30.

Noticeably, neither of the two authors seem to be familiar with the precession of the equinoxes. Because of this phenomenon, the declination of a star changes slowly over centuries, as well as the angular distance measured from the celestial pole.

At the end of the 2nd millennium BC, Polaris (α UMi) was not located near the north celestial pole, as shown in Fig.1. Actually, it was the farthest star of its constellation from the north pole, being as far away as the brighter ϵ UMa. At that time, the two constellations lay almost at the same angular distance from the pole, being the bright stars of Ursa Major stars closer to it than in the present time.

The situation changed little in Alexandrian times (around 300 BC), when Ursa Minor stars were about 10° away from the north pole, being Polaris closer by only 4° to it, while Ursa Major stars were about 20° away. Then, Aratus might have been aware that Ursa Minor revolved entirely in a smaller circle than Ursa Major did.

Hence, the question raised by Finkelberg (2004) poses a false problem, because it is not necessary to assume that «ἥ τ' αὐτοῦ στρέφεται» can refer only to Ursa Minor and to conclude that these lines were a later interpolation. In fact, both constellations appeared to 'turn about in the same spot' in the Homeric age, but it was definitely the most conspicuous one, Ursa Major, to be used as a skymark for navigation, as attested by Homer's epithet ἑλίκωπες Ἀχαιοί.

As a further analysis, the technical term ποντοπορεύω, 'to pass over the sea', present in an astronomical context allows us to determine if celestial navigation was the result of transmission of advanced maritime knowledge from earlier ages. Albeit some scholars were skeptical about this, Tartaron (2013) pointed out that stellar navigation, as described in Homer, must also have been familiar to Mycenaeans to maintain the course, when their ships encountered extended periods in the open sea.

There are compelling arguments in favour of Aegean Bronze Age celestial navigation. On one hand, some Early Cycladic frying pans carry incised depictions of symbols related to sea (waves), sky (sun/stars), and maritime travels (longboats). This archaeological evidence is suggestive of celestial navigation (Agouridis 1997). On the other hand, the proliferation of coastal settlements and long-distance maritime relations throughout the Aegean necessarily involved direct nighttime journeys, quicker than coastal routes. Therefore, Tartaron (2013) concluded that, almost certainly, stellar navigation was already practised during the pre-Homeric period.

3.2 The Theoclymenus' Vision

At the end of book XX, on Apollo's feast day the seer Theoclymenus has an eerie vision during the banquet and foretells the imminent doom of the Suitors, while they are celebrating (*Od.* 20, 350–357):

τοῖσι δὲ καὶ μετέειπε Θεοκλύμενος θεοειδής·
«ἆ δειλοί, τί κακὸν τόδε πάσχετε; νυκτὶ μὲν ὑμέων
εἰλύαται κεφαλαί τε πρόσωπά τε νέρθε τε γοῦνα·
οἰμωγὴ δὲ δέδηε, δεδάκρυνται δὲ παρειαί·

αἵματι δ' ἐρράδαται τοῖχοι καλαί τε μεσόδμαι·
εἰδώλων δὲ πλέον πρόθυρον, πλείη δὲ καὶ αὐλή,
ἱεμένων Ἔρεβόσδε ὑπὸ ζόφον· ἠέλιος δὲ
οὐρανοῦ ἐξαπόλωλε, κακὴ δ' ἐπιδέδρομεν ἀχλύς.»

Godlike Theoclymenus now spoke out among them:
≪Poor wretches, what evil has come on you? Your heads and faces
and the knees underneath you are shrouded in night and darkness;
a sound of wailing has broken out, your cheeks are covered
with tears, and the walls bleed, and the fine supporting pillars.
All the forecourt is huddled with ghosts, the yard is full of them
as they flock down to the underworld and the darkness. The sun
has perished out of the sky, and a foul mist has come over.≫[28]

Some ancient commentators, among them Plutarch and Heraclitus the Allegorist, believed this passage to be a poetic description of a total solar eclipse.[29] This opinion has been further confirmed by *P. Oxy.* 3710, which preserves a fragment of an earlier exegetical commentary on the *Odyssey* (Haslam 1986). All these speculations rely on the fact that it was new Moon.[30] On the other hand, the scholia to the *Odyssey* mention this astronomical interpretation of the Theoclymenus' prophecy, but the scholiasts reject this hypothesis, observing that the speech refers only to the Suitors who will not see the sunlight, i.e., who are going to perish.[31]

According to Dodds (1951), the symbolic vision of the seer belongs to prophetic madness, being a manifestation of Greek irrationality. This interpretation is followed by the majority of modern scholars, taking the passage as a metaphor.[32]

However, the idea that this passage alludes to a real solar eclipse has been revived in modern times. Under this assumption, Schoch (1926) proved that over a long period only the solar eclipse of 16 April 1178 BC was total over the Ionian Islands, arguing a confirmation of the traditional date of the fall of Troy, ten years before. Dörpfeld (1928) firmly rejected the argument, whereas Shewan (1928) did not exclude that an eclipse is meant in the passage, but found difficulties with the date, since the action takes place in autumn. Conversely, asserting that Odysseus' journey home takes place in spring, MacDonald (1967) found the theory attractive.

Baikouzis and Magnasco (2008) have recently renewed the question, with different arguments. They have considered several astronomical references in the *Odyssey*:
(a) the simultaneous presence of the Pleiades and Boötes in spring (*Od.* 5, 272–275);
(b) Venus rising before dawn, five days before Odysseus' return (*Od.* 13, 93–96);
(c) the new Moon (*Od.* 14, 161–162 = 19, 306–307).
Furthermore, they have *conjectured* that Hermes' travel westwards to Ogygia and

---

[28] The translation is taken from Lattimore (cit.).

[29] Plu. *De facie*, 931e-f; Heraclit. *Quaestiones Homericae*, 75; see also [Plu.] *Vit. Hom.* 2, 108.

[30] *Od.* 14, 161–162 = *Od.* 19, 306–307. The return of the hero on the day of the new Moon was an important element of the narrative, cf. Heubeck and Hoekstra (1989), ad loc.; see also Austin (1975).

[31] See, e.g., Eust. *ad Od.* 1895, 1–17.

[32] Cf. Russo et al. (1992) ad loc.; see also Broggiato (2003).

return, as told in book V, represents the retrograde motion of Mercury. In the range 1250-1115 BC, the single date of 16 April 1178 BC matches all these astronomical events, occurring simultaneously, independently of the disputed eclipse reference. Thus, they speculate that these references may indeed refer to that historical eclipse, and may be structural, defining an astronomical timeline of the Homeric epic.

Gainsford (2012) has strongly argued against this conclusion. There is no proof in any ancient Greek texts that the gods' movements, like Hermes' travel, represent the movements of heavenly bodies. Moreover, many passages point to an autumn setting. On the latter basis, Papamarinopoulos et al. (2012) have analysed again the astronomical data mentioned by Homer, suggesting that the annular solar eclipse of 30 October 1207 BC is the phenomenon described in the Theoclymenus' vision.

In our view, the passage might represent the surviving memory of the impressive darkness experienced during eclipses, which occurred over centuries, then transmitted *per ora virum*, rather than referring to an actual event.[33]

4 Concluding Remarks

In this work, we have reviewed the current status of research on the astronomical knowledge in Homer, summarising recent advances. In spite of earlier qualms, scholars have indeed started to appreciate the potential of contextual interpretations of references to heavenly bodies and phenomena in the Homeric poems. However, although a certain amount of this information can be found in monographs and commentaries,[34] a broad archaeoastronomical perspective on the subject is still lacking.

We have shown that linguistic and philological investigations, supported by archaeological and astronomical data, indicate that the perception of celestial events played a role in the daily life of ancient Greeks in pre-Hellenic and Homeric times. Moreover, many arguments suggest that the knowledge and understanding of astronomical phenomena were at a higher level than previously thought.

We have presented two case studies concerning the *Odyssey*. We have clearly demonstrated that contradictions arise when the analysis of literary texts relies on a single approach, either philological or astronomical, and neglects other evidence.

In the future, we aim to extend our analysis of the Homeric texts. In the *Odyssey*, at least two other passages deserve attention: the reference to the new Moon,[35] containing the puzzling term λυκάβας, and the mention of Venus,[36] which we have touched on above (see Sect. 3.2). Furthermore, it would be worth investigating the

[33] A similar opinion is expressed by Shewan (1928); Dicks (1970); Theodossiou et al. (2011); Revello (2013).

[34] See, e.g.: *Iliad*: Kirk, G.S., ed.: The Iliad: A Commentary. 6 voll. (Cambridge, 1985–1993); *Odyssey*: Heubeck, A., et al.: A Commentary on Homer's Odyssey. 3 voll. (Oxford 1988–1992).

[35] *Od.* 14, 161–162 = 19, 306–307 (see above).

[36] *Od.* 13, 93–96.

relationship between the imagery of astronomical events in the Homeric poems and in Archaic lyric poetry.

Acknowledgements The authors wish to thank Prof. P.M. Militello for useful references and Dr. G. Cutispoto for reading the manuscript. Financial support from the Dipartimento di Fisica e Astronomia of the Università degli Studi di Catania is gratefully acknowledged.

References

Aaboe, A.: Scientific astronomy in antiquity. Philos. Trans. R. Soc. Lond. A **276**(1257), 21–42 (1974)

Aaboe, A.: Episodes from the Early History of Astronomy. Springer, Berlin (2001)

Agouridis, C.: Sea routes and navigation in the third millennium Aegean. Oxf. J. Archaeol. **16**(1), 1–24 (1997)

Austin, N.: Archery at the Dark of the Moon. Poetic Problems in Homer's Odyssey. University of California Press, Berkeley (1975)

Aveni, A.F.: L'archeoastronomia: scopi, ricerche, risultati. G. Astron. **13**(3), 15–18 (1987)

Aveni, A.F.: Introduction: whither archaeoastronomy?. In: Aveni, A.F. (ed.) World Archaeoastronomy. Selected Papers from the 2nd Oxford International Conference on Archaeoastronomy, pp. 3–12. Cambridge University Press, Cambridge (1989)

Aveni, A., Romano, G.: Temple Orientations in Magna Graecia and Sicily. J. Hist. Astron. Suppl. **31**, S51–S57 (2000)

Baikouzis, C., Magnasco, M.O.: Is an eclipse described in the Odyssey? Proc. Natl. Acad. Sci. **105**, 8823–8828 (2008)

Beekes, R.S.P.: ἔτος and ἐνιαυτός in Homeric formulae. Glotta **LXVII**, 138–143 (1969)

Boutsikas, E., Ruggles, C.L.N.: Temples, stars, and ritual landscapes: the potential for archaeoastronomy in ancient greece. Am. J. Archaeol. **115**(1), 55–68 (2011)

Broggiato, M.: Cratete di Mallo: i frammenti. In: Spezia, L. (ed.) Agorà (2001)

Broggiato, M.: Interpretazioni antiche e moderne della visione di Teoclimeno nell'<<Odissea>> (Od. 20, 351–357). ῾Ρυσμός: studi di poesia, metrica e musica greca offerti dagli allievi a Luigi Enrico Rossi per i suoi settant'anni. Quaderni dei Seminari Romani di Cultura Greca, 6, pp. 63–72. Quasar, Rome (2003)

Carnevale, F., Monaco, M., Ranieri, M.: A study on the orientation of Greek theatres. In: Orlando, A. (ed.) The Light, the Stones and the Sacred, Astrophysics and Space Science Proceedings, **48**, 107–121 (2017)

Catalin, A.: Watching for Orion: a note on Od. 5.274 = Il. 18.488. Class. Q. **60**(1), 250–253 (2010)

Chadwick, J.: The Mycenaean World. Cambridge University Press, Cambridge (1976)

Chantraine, P.: Grammaire homérique, I: Phonétique et morphologie. Klincksieck, Paris (1942)

Chantraine, P.: Dictionnaire étymologique de la langue grecque: histoire des mots. Klincksieck, Paris (1968)

Cline, E.H.: The Oxford handbook of the Bronze Age Aegean (ca. 3000–1000 BC). Oxford University Press, Oxford (2010)

Dettori, E.: Arat. 469–476: una risposta a Il. 8,555-559?. Rhein. Mus. **151**(3–4), 426–429 (2008)

Dicks, D.R.: Early Greek Astronomy to Aristotle. Cornell University Press, Ithaca, N.Y. (1970)

Dodds, E.R.: The Greeks and the Irrational. University of California Press, Berkeley (1951)

Dörpfeld, W.: Die Datierung des trojanischen Krieges. Die Sterne **16**, 186–187 (1928)

Edwards, M.W.: The Iliad: A Commentary, vol. 5, Books 17–20. Cambridge University Press, Cambridge (1991)

Emlyn-Jones, C.: ἔτος and ἐνιαυτός in Homeric formulae. Glotta **XLV**, 156–161 (1967)

Evans, J.: The History and Practice of Ancient Astronomy. Oxford University Press, Oxford (1998)

Finkelberg, M.: <<She turns about in the same spot and watches for Orion>>: ancient criticism and exegesis of Od. 5.274 = Il. 18.488. Greek, Roman and Byzantine Stud. **44**(3), 231–244 (2004)
Fisher, R.: Astronomy and the calendar in Hesiod. Échos du monde classique = Class. Views **21**, 58–63 (1977)
Gainsford, P.: Odyssey 20.356–57 and the Eclipse of 1178 b.c.e.: a response to Baikouzis and Magnasco. Trans. Am. Philol. Assoc. **142**(1), 1–22 (2012)
Gendler, J.R.: Jupiter-Saturn conjunctions in Homer's Odyssey and Iliad. Bull. Am. Astron. Soc. **16**, 489–490 (1984)
Grey, S.: Astronomical knowledge in Homer's Odyssey: book IX case study. In: Kotarcic, A., Underwood, D. (eds.) Proceedings of the Annual Meeting of Postgraduates in Ancient Literature 2013. University St Andrews, St Andrews (2013)
Hannah, R.: The constellations on Achilles' shield (Iliad 18. 485–489). Electron. Antiq. **2**(4) (1994)
Hannah, R.: Odysseus's navigation. Prudentia **29**(1), 15–33 (1997)
Hannah, R., Magli, G., Orlando, A.: New Insights on the Akragas' Complex of Demeter and Persephone: The Role of the Moon. In: Orlando, A. (ed.) The Light, the Stones and the Sacred, Astrophysics and Space Science Proceedings, **48**, 95–106 (2017)
Haslam, M.W.: The Oxyrhynchus Papyri, LIII. Egypt Exploration Society, London (1986)
Havelock, E.A.: Preface to Plato. Harvard University Press, Cambridge (1963)
Heath, T.L.: Greek Astronomy. Courier Dover Publications, London (1932)
Herbert, S.C.: The Orientation of Greek Temples. Palest. Explor. Q. **116**(1), 15–18 (1984)
Heubeck, A., Hoekstra, A.: A Commentary on Homer's Odyssey, vol. 2, Books 9–16. Oxford University Press, Oxford (1989)
Heubeck, A., West, S., Hainsworth, J.B.: A Commentary on Homer's Odyssey, vol. 1, Books 1–8. Oxford University Press, Oxford (1988)
MacDonald, T.L.: The season of the Odyssey. J. Br. Astron. Assoc. **77**, 324–327 (1967)
Murray, G.: The Rise of the Greek Epic. Clarendon Press, Oxford (1924)
Neugebauer, O.: A History of Ancient Mathematical Astronomy. Part 1; Part 2; Part 3. Springer, Berlin (1975)
North, J.: Cosmos: An Illustrated History of Astronomy and Cosmology. University Press of Chicago, Chicago (2008)
Papamarinopoulos, S.P., Preka-Papadema, P., Antonopoulos, P., Mitropetrou, H., Tsironi, A., Mitropetros, P.: A new astronomical dating of Odysseus return to Ithaca. Mediterr. Archaeol. Archaeom. **12**, 117–128 (2012)
Papamarinopoulos, S.P., Preka-Papadema, P., Antonopoulos, P., Mitropetrou, H., Tsironi, A., Mitropetros, P.: The Anatomy of a complex Astronomical Phenomenon described in the Odyssey. Mediterr. Archaeol. Archaeom. **13**, 69–82 (2013)
Parker, R.A.: Ancient Egyptian astronomy. Phil. Trans. R. Soc. Lond. A **276**(1257), 51–65 (1974)
Pasquali, G.: Storia della Tradizione e Critica del Testo. Le Monnier, Florence (1934)
Phillips, J.H.: The constellations on Achilles' shield (Iliad 18, 485–489). Liverp. Class. Mon. **V**, 179–180 (1980)
de Price, D.J.: Solla: Science Since Babylon. Yale University Press, New Haven (1975)
Renaud, J.-M.: Le catastérisme d'Orion. In: Bakhouche, B., Moreau, A.M., Turpin, J.-C. (eds.) Les astres: actes du colloque international de Montpellier, 23-25 mars 1995, Séminaire d'étude des mentalités antiques, vol. 1, pp. 83–93. Publications de la recherche, Univ. Paul-Valéry, Montpellier (1996)
Renaud, J.-M.: Le catastérisme chez Homère: le cas d'Orion. Gaia **7**, 205–214 (2003)
Renaud, J.-M.: Le mythe d'Orion: sa signification, sa place parmi les autres mythes grecs et son apport à la connaissance de la mentalité antique. Centre Informatique de Philosophie et Lettres. University of Liège, Liège (2004a)
Renaud, J.-M.: Monde sauvage et monde civilisé dans le mythe: le cas d'Orion. In: Charpentier, M.-C. (ed.) Les espaces du sauvage dans le monde antique: approches et définitions, pp. 279–290. Les Belles Lettres, Paris (2004b)
Revello, M.: Sole, luna ed eclissi in Omero. Technai **4**, 13–32 (2013)

Richardson, N.J.: The Iliad: a Commentary, vol. 6, Books 21–24. Cambridge University Press, Cambridge (1993)

Ropars, J.-M.: Une signification de l'<<Odyssée>>. Bull. Assoc. Guillaume Budé **1**(1), 82–95 (2003)

Ruggles, C.L.N.: Megalithic astronomy—the last five years. Vistas Astron. **27**, 231–289 (1984)

Ruggles, C.L.N.: Handbook of Archaeoastronomy and Ethnoastronomy. 3 voll. Springer, New York (2015)

Ruggles, C.L.N., Urton, G.: Introduction. In: Ruggles, C.L.N., Urton, G. (eds.) Skywatching in the Ancient World. New Perspectives in Cultural Astronomy, pp. 1–16. University Press of Colorado, Boulder (2006)

Russo, J., Fernández-Galiano, M., Heubeck, A.: A Commentary on Homer's Odyssey vol. 3, Books 17–24. Oxford University Press, Oxford (1992)

Salt, A.M.: The astronomical Orientation of ancient Greek Temples. PLoS One (2009). doi:10.1371/journal.pone.0007903

Schoch, C.: The eclipse of Odysseus. Observatory **49**, 19–21 (1926)

Shewan, A.: Two ancient eclipses. Class. World **XXI**, 196–200 (1928)

Tartaron, T.F.: Maritime Networks in the Mycenaean World. Cambridge University Press, Cambridge (2013)

Theodossiou, E., Manimanis, V.N., Mantarakis, P., Dimitrijevic, M.S.: Astronomy and Constellations in the Iliad and Odyssey. J. Astron. Hist. Herit. **14**, 22–30 (2011)

Wood, F., Wood, K.: Homer's Secret Odyssey. History Press ltd., Stroud, Gloucestershire (2011)

Ratio Siderum in Pliny the Elder: Pleiades, Light and Wheat

Lucia Bonacci

Abstract The story of the “seven sisters”, the brightest star visible to the naked eye and forming the Pleiades constellation, has always fascinated cultures, more or less close to us, for the mythological stories related to them and for the influence they have had in the construction of temples of the ancient world (to name a few examples, the temple of Asklepios at Epidaurus and that of Bacchus in Athens), oriented according to their rise or their set. As their importance in the Roman world? In this regard, this article will be highlight, through the eighteenth book of Pliny the Elder dedicated to farm work, the *ratio siderum* of this constellation and its relationship with the light and the sacred, especially in relation to the cereal growing.

1 Introduction

In what is conventionally referred to as the second part of XVIII book, Pliny writes very interesting pages on the influence of stars and constellations in the cereal growing. In fact, he had not been the first to evidence this close connection, since, for example, already Hesiod and Virgil have stressed the importance of the binomial heaven/earth.

The Pleiades, visible in the constellation Taurus, mark the times of harvest and plowing and their heliacal rising occurred around the sixth day before the Ides of May. With their full presence in the sky, fireflies appear on earth: this indicates that the barley and the wheat is ripe in the fields and that the summer sowing of millet and panic can be undertaken.

The firefly brings an explicit reference to the splendor, to the fire and to the light, that even back in the etymology of the same word. This, combined with the words of the Latin scholar, will give way to observe how this constellation is by the combination of light, earth and the sacred, through the cultivation of cereals.

L. Bonacci (✉)
Università degli Studi della Calabria, Rende, Italy
e-mail: lucia.bonacci@unical.it

A. Orlando (ed.), *The Light, The Stones and The Sacred*,
Astrophysics and Space Science Proceedings 48,
DOI 10.1007/978-3-319-54487-8_11

1.1 Mythology About Grain

The mysterious ear of corn, therefore, takes the divine nourishment: the food given by Demeter to Triptolemus, as a reward for the revelation of the captor of his daughter Core/Persephone. Yet, in Egypt, the ear was associated with the myth of Isis and Osiris, whose body's part were scattered in the world by cruel brother Seth, and recovered from the sister and wife Isis. The god, resurrected in the apotheosis of immortality, became Ra, Lord of the Underworld, patron of the resurrection and, consequently, the vegetation.

As in the case of the Egyptian myth, also in the Babylonian mythology, Tammuz, son of Ishtar, becomes the spirit of wheat, being a young man who died each year to be reborn to new life, just as the seed that is planted to become shoot in spring.

In ancient Rome, goddess of the harvest was Ceres, whose Latin name, deriving from the root of the inchoative *cresco* and of the causative *creare*, personifies Growth. The blonde goddess, such as wheat, were named even the ears, *cerealis.*[1]

The Greek goddess was confused with the Roman divinity, to the point that it also inherited her daughter, that within Latin became Persephone, and the myth of his abduction.

Founded in the fifth century BC, a sanctuary of the goddess was also placed at the Aventine and in her honor were celebrated the *Cerealia*, every April 12th: during these celebrations were sacrificed oxen and pigs and offered fruits and honey. Always on the Aventine, it remembered a construction and a *dedicatio* of a temple to *Ceres*, *Liber et Libera*, desired by the dictator *Postumius*, between 495 and 493 BC.

[1]The legend says that, near Enna, Ceres fertilized the land, to bring life to the playfulness of his gifts. Ceres, sister of Jupiter, was venerated as a goddess who had taught men to till the fields and make them lush. Ceres was the Mother Earth who had a lovely daughter named Persephone. The Romans explained the turning of the seasons with the following story: Ceres was the sister of Jupiter, and Proserpine was her daughter. Proserpine was kidnapped by Pluto, god of the underworld, to be his bride. By the time, Ceres followed her daughter, she was gone into the earth. Making matters worse, Ceres learned that Pluto had been given Jupiter's approval to be the husband of her daughter. Ceres was so angry that she went to live in the world of men, disguised as an old woman, and stopped all the plants and crops from growing, causing a famine. Jupiter and the other gods tried to get her to change her mind but she was adamant. Jupiter eventually realized that he had to get Proserpine back from the underworld, and sent for her. Unfortunately, Pluto secretly gave his food before she left, and once one had eaten in the underworld one could not forever leave. Proserpine was therefore forced to return to the underworld for four months every year. She comes out in spring and spends the time until autumn with Ceres, but has to go back to the underworld in the winter. Her parting from Ceres every fall is why plants lose their leaves, seeds lie dormant under the ground, dormant under the ground, and nothing grows until spring when Proserpine is reunited with her mother.

The spelt, then, heavily consumed by the Roman people, it was sacred to the point of pushing Numa to establish a celebration, the *fornacalia*, which lasted no more than February 17th.

The name of the party, as we learn from Alfredo Cattabiani, stemmed from *fornax*, furnace, which was also the name of a goddess and this was not just an appointment for roasting the barley, but had a political purpose, since it provided for the meeting of all members of the same curia in a hole indicated by *tabellae.*[2]

Spelt was protected by the Vestal, which, until the end of the fourth century AD, did not cease to prepare for the cult the *mola salsa*, spelt mixed with salt both cooked and uncooked, said *muries* and essential for blood sacrifices.[3]

1.2 The Sky of the Ancients

Mythology and rituals related to grain and agricultural activities originate even more remote than the reconstruction exposed. These aspects, in fact, already in prehistoric times, are connected with the observation and the mysteries of the starry sky.

At Dalby, in Denmark, was found a stele, representing some constellations (Virgo, Leo, Cancer, Gemini, Auriga, just to mention the most famous) (Fig. 1) in the position where they would incur around May 20th, that is the period related to spring agricultural activities. Moreover, in a community with predominantly agricultural economy, it probably thought important to set in a stone of the most significant moments of the seasonal cycle.

The importance of the binomial air/land was, on several occasions, shown already by Hesiod and Virgil. Later, Pliny the Elder devotes an entire book (the XVIII) of his *Naturalis Historia*, to the quality of the cereal products, to the climate and soils best suited to house them; he also talks about the regions most producers, not follow, however, a strictly linear discussion, because he opens to technical and antiquarian digressions, flowing in moralistic considerations. In what is conventionally referred to as the second part of this book, Pliny writes extremely interesting pages on the influence of the position of the stars and constellations in the growth of cereals.

The purpose of this article, which rightly can fit in the general field of the historiography of astronomy, is to track, analyze and examine the Pliny's considerations on stars and their influences on cereals.

[2]Cattabiani (2010), p. 446; Ovid., Fas., II, 527–529.

[3]Cattabiani (2010), p. 447; Festo, 152, L.

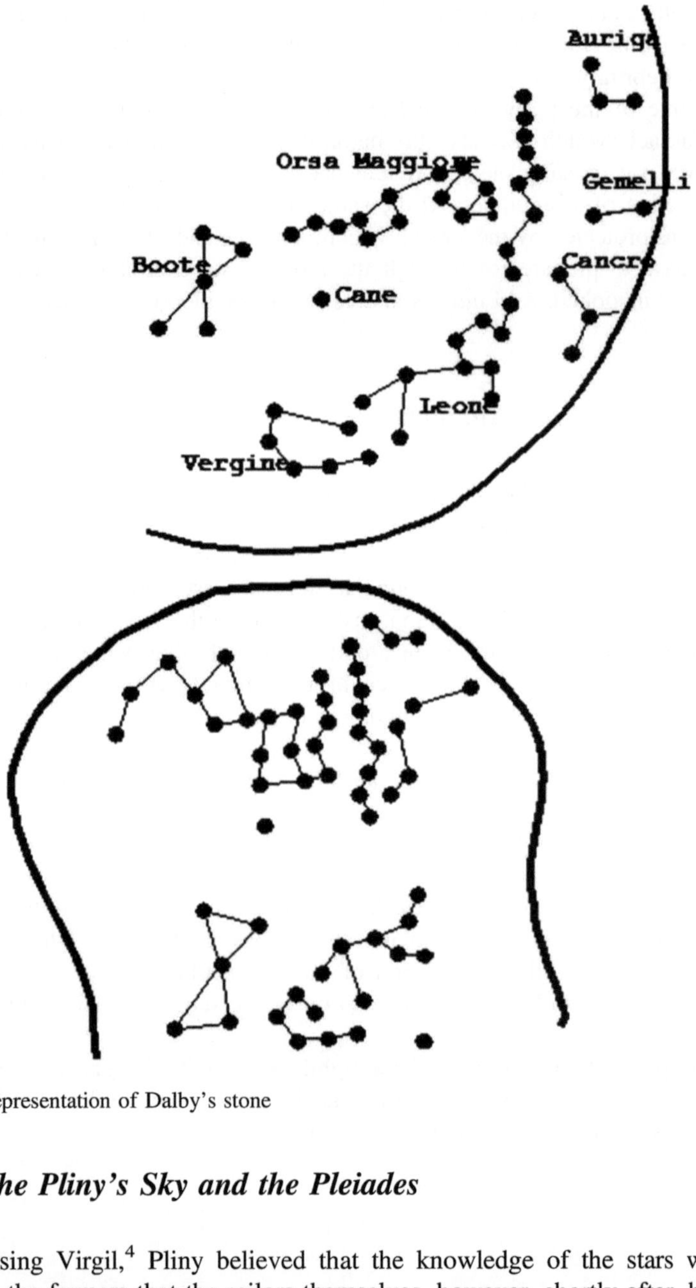

Fig. 1 Representation of Dalby's stone

1.3 The Pliny's Sky and the Pleiades

Paraphrasing Virgil,[4] Pliny believed that the knowledge of the stars was more useful to the farmers that the sailors themselves, however, shortly after, he openly manifests his skepticism about the possibility of predicting the future by reading

[4]Virg., Georg., I, 204 ff.

stars.[5] Pliny introduces this close relationship between heaven and earth already in the second book of his work, dedicated to cosmology, in which the earth, that generates and nourishes us, is defined *maternae venerationis.*[6]

The Pliny's calculation and observation are based on the dictates of the calendar reform, introduced by Caesar in 46 BC and created following the advice of the astronomer Sosigenes. Specifically, Pliny will list the theories of different scholars, Latin and Greek, about the cultivation of cereals and their relationship with the sky.

The Latin scholar, explaining the stars and the best time to cultivate the land, divides the treatment into regions, preferring the Italy. The rest of ecumene will be divided as follows: Attica (also including the Cyclades), Macedonia (with Magnesia and Thrace), Egypt (here, he will also address Phoenicia, Cyprus and Cilicia), Boeotia (with Locris and Phocis) the Hellespont to Mount Athos, the Ionia (where cluster Asia Minor and the islands), the Peloponnese (Achaia and along with the lands to the west) and the Chaldeans, meaning, with the latter, Assyria and Babylon. Pliny not includes Africa, Spain and Gaul, since, in these territories, no one has ever made careful to observe the stars.

Specifically, Pliny begins his disquisition on wheat, tied to the rising of the Pleiades, as already underlined Virgil, calling them *Eoae Atlantides*, or "the morning".[7]

The story of the "seven sisters", the brightest stars visible to the naked eye and making up the constellation of the Pleiades, has always fascinated many cultures more or less close to us, to the point that their influence can be glimpsed in the construction of temples and buildings of the ancient world (just to mention a few examples, the temple of Asklepios at Epidaurus and that of Bacchus in Athens), oriented according to their rise or their set.

According to mythology, the seven stars are Maia, Electra, Taygeta, Alcyone, Celeo, Sterope and Merope: already Hesiod (*The Works and the Days*, 383) defines them daughters of Atlas, and Hellanicus talked a lot about them in his work *Atlantis*.

---

[5] *Spes ardua, immensa, misceri posse caelestem divinitatem inperitiae rusticae, sed temptanda tam grandi vitae emolumento. Prius tamen sideralis difficultas, quam sensere etiam periti, subicienda contemplationi est, quo deinde laetior mens discedat a caelo et facta sentiat, quae futura praenosci non possint.* It is an arduous attempt, and almost beyond all hope of success, to make an endeavour to introduce the divine science of the heavens to the uninformed mind of the rustic; still, however, with a view to such vast practical results as must be derived from this kind of knowledge, I shall make the attempt. There are some astronomical difficulties, however, which have been experienced by the learned even, that ought to be first submitted for consideration, in order that the mind may feel some encouragement on abandoning the study of the heavens, and may be acquainted with facts at least, even though it is still unable to see into futurity [Plin. N.H. XVIII, 56, 206].

[6] See Plin. N.H. II, 63, 154.

[7] In detail, Pliny means, taking the words of Virgil, that the Pleiades, in the sky of Rome, fade away in the morning. Their sunset, or their disappearance in the heavenly vision, is placed at the beginning of November.

Their mother had been Pleione, daughter of Ocean, who had given birth on Mount Cyllene, in Arcadia; so they were considered at the same time celestial nymphs and mountains nymphs.

On their migration from the mountains to the sky are given two reasons: according to the first (tale by Aeschylus), it would be reason sorrow for the misfortunes of his father Atlas; according to others (tale by Pindar), stocks and chase, with his mother, to the giant Orion had fled before him for five years, until Jupiter had moved to heaven the whole group, including the pursuer with his dog.

1.4 The Pleiades and the Agricultural Calendar

The Pleiades are visible in Taurus constellation and rise around the sixth day before the Ides of May (around May 10th), about to set on the third day before the Ides of November (November 11th). Hesiod already affirmed that, with the appearance in the sky of the Pleiades, began the harvest, while the plowing at their sunset, placing his attention to the Boeotia:

αἳ δή τοι νύκτας τε καὶ ἤματα τεσσαράκοντα κεκρύφαται, αὖτις δὲ περιπλομένου ἐνιαυτοῦ φαίνονται τὰ πρῶτα χαρασσομένοιο σιδήρου.[8]

Pliny, before moving on his remarks and his advice on best astral season to proceed with the cereal, reviews the main theories on the subject. Virgil, for example, may want to sow *triticum* and spelt after sunset of the Seven Sisters; others, however, to sow them before their sunset, especially in arid and hot areas, such as the seed is preserved and the first rain will sprout them quickly.[9]

Pliny writes that the sowing of wheat is done in parallel with the early sunset of the Pleiades and this action, since many farmers, is anticipated eleventh day after the fall equinox, with the rise of the Crown that determines the rains.[10]

According to Pliny, the true signal that starts the operation of sowing is not the rain, but the falling of the leaves which occurs, again, in relation to the Pleiades, and more precisely at their sunset.

> hoc ipso vergiliarum occasu fieri putant aliquid a.d. III idus Novembris, ut diximus, servatque id sidus etiam vestis institores, et est in caelo notatu facillimum.[11]

The presence of wind, called foehn, requires continued commitment of farmers in many different activities, including the polishing of the wheat fields and the

---

[8]"Forty nights and days they are hidden and appear again as the year moves round, when first you sharpen your sickle" (Hes. WD, 383–386).

[9]See Plin. N.H. XVIII, 56, 202–203; Virg., Georg., I, 219–221; Colum., II, 8, 4.

[10]Crown arose 4–15 of October and it was connected to the rains. See Colum., XI, 2, 73.

[11]"The true method to be adopted, however, is not to sow until the leaves begin to fall. Some persons are of opinion that this takes place at the setting of the Vergiliæ, or the third day before the ides of November" (Plin. N.H. XVIII, 60, 225).

weeding of grains, especially spelt; barley is weeding, however, when the weather is dry and the pruning will finish before the spring equinox.[12]

With the full presence of the Pleiades in the sky, fireflies appear on earth; this indicates that the barley and the wheat are ripe in the fields and that can be undertaken the summer sowing of millet and panic.

Firefly brings an explicit reference to the splendor, to the fire and to the light, that also returns in the etymology of the word itself.

The fireflies were called *cicindelae* by the Latins, lemma with doubling, from the root *cand-* (ignite, burn), the same one that is also in the verb *incendo* (fire preventing).

Similarly the Greek term—*lampyrides*—is etymologically connect to lightning, with the meaning to *risplendo*, shine.[13]

2 The Pleiades: The Light and the Stone

As previously mentioned, Pleiades seem to have oriented the construction of buildings and sacred sites, as well as being represented on various findings, discovered in various archaeological campaigns.

Indeed, there is much debate about the authenticity of these discoveries, or more specifically, about the real existence of a connection with this constellation.

For example, in Tikal there is the first pyramid city of the Maya. Some historians have concluded that its disposal was created in veneration of the cosmic Pleaides.[14] In fact, the Maya obsessed in the Pleiades because this stars measure seasons to plant and harvest corn.

The importance of the Pleiades is also attested in earlier periods than those reported so far. In fact, the Babylonians began the new year at the heliacal rise of the Pleiades, just like did their ancestors Sumerians.[15]

They depict them on the boundary stones and had in high regard the star Alcyone (Fig. 2). Pleiades were the subject of a cult: it is said that the idol Succoth-Benoth was represented by a hen with chicks, and thus going up to a linguistic variant, always used to distinguish this constellation.[16]

Although hypotheses must be highly controversial, it is important to remember the discovery of the Nebra disk (Fig. 3), a bronze object dated 1600 BC. The incisions in gold on the field of large Disk represent another lower disk and a half

[12]See Plin. N.H. XVIII, 65, 240–242; Colum., II, 11, 5.

[13]An old Italian proverb states: "Giugno lucciolaio festa nel granaio". In other words, if in June there will be many fireflies, harvest will be plentiful.

[14]The opinions on this issue are conflicting (see e.g.: Aveni 1975, pp. 163–190; Aveni 1977, pp. 165–202; Aveni 1980; Coe 1975, pp. 3–31.

[15]See Ridpath (1994), p. 11.

[16]See Mulas-Sanna (2013), p. 22.

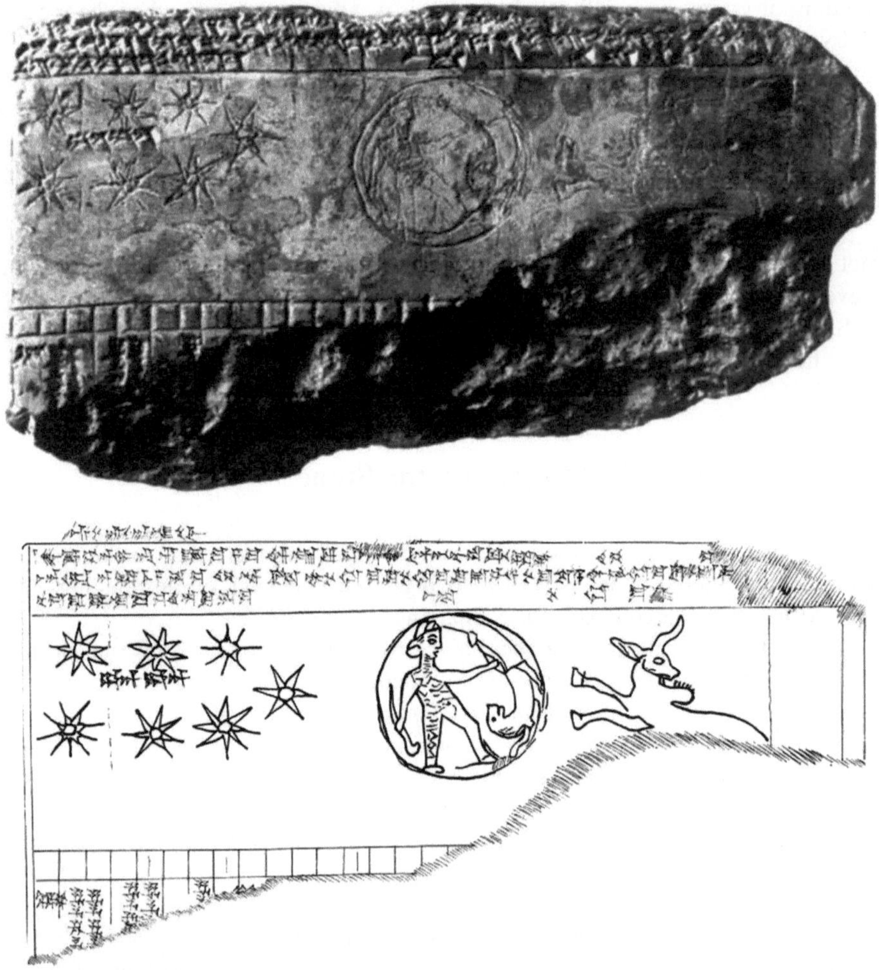

Fig. 2 Late Babylonian tablet that represents the moon exalted in Taurus. The 3 symbols illustrate the constellations *MUL.MUL zappu* "The Star Cluster" (Greek: Pleiades), the moon, and *MULGU4.AN.NA alû/is lê* "The Steer/Bull of Heaven" (Greek: Taurus) in the order of the moon's transit. The moon in between is positioned to show its exaltation in Taurus. In the lunar disk the God Marduk is wrestling a lion eclipse (Staatliche Museen zu Berlin, no. VAT 7851)

moon; two arcs are located at the opposite edges of the large disk and in the end of an arc of a circle. The field of the disc is then scattered by 30 tiny discs called circles or points. We don't know if this was an ornamental object, an magical ritual object or another. It's certainly an object of value, as forged in gold.

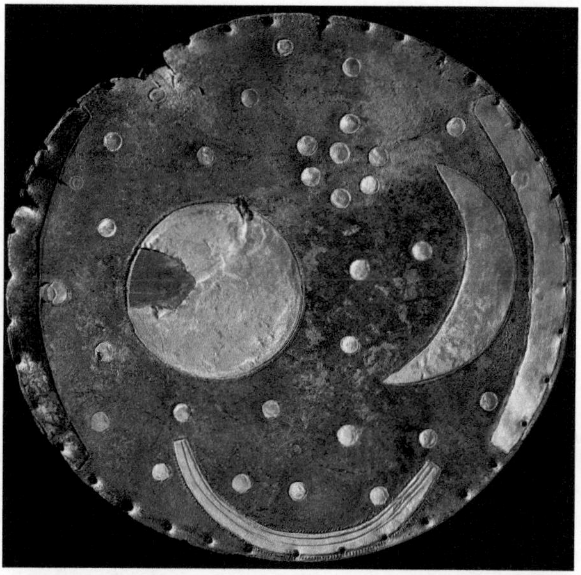

Fig. 3 Nebra sky disk (Halle State Museum of Prehistory, Halle (Saale), Germany)

According to some scholars, the disk represents a geometric-mathematical vision of the cosmos, superior to ours.[17] Long discussions have been discussed on the interpretation of the seven little circles.

The promoter of the hypothesis that sees in them the depiction of the Pleiades is Wolfhard Schlosser. These circles could also represent Orion, this one located on Axis of Heaven as the Pleiades.[18]

According to others, the representation of the Pleiades is not realistic than the actual configuration in the sky, despite the relative angular dimensions between those of small asterism and those of the Sun and the Moon correspond well enough.[19] The possibility that the small cluster of stars was Pleiades it can explain as this is the observation of a group of stars of small size and low light, it's easy to see the Pleiades.[20]

Finally and going back to the ancient world, we remind that the disk will be in perfect accord with some verses of the XVIII book of the Iliad, in which the blind poet explains the astronomical decorations of the Achilles' shield.

> Therein he wrought the earth, therein the heavens therein the sea,/ and the unwearied sun, and the moon at the full/, and therein all the constellations wherewith heaven is crowned/the Pleiades, and the Hyades and the mighty Orion.[21]

[17]See Droy (2004), pp. 57–58.
[18]See Droy (2004), p. 57.
[19]See Gaspani (2009), pp. 32–39.
[20]See Gaspani (2008), pp. 32–37; id. 2009, pp. 32–39.
[21]Hom. Il. 18, 483–486.

Light is symbolized by the gold fields that, generally after the Solstice, are subject to the harvest: the Light on Earth has turned in a sacred and alchemical manner in wheat that is Life. The cycle and cosmic dualism of light and darkness celebrates, once again and always in all religions, the tragedy of Life and Death.

In parallel, in the sky, from the winter solstice the days gradually grow and the Sun grows up until the triumphant apotheosis at the beginning of the summer season, with the entry in the maternal sign of Cancer. From here, it begins a slow decline that brings a transfer of energy to the underworld.

In the darkness it's alive the seed of Light, as the Light lives in the Darkness. Similarly, in the Light there is Holiness and there are also the splendor of the sky and the solidity of the Stone and the Earth.

3 Conclusions

The symbolism related to the Light and the Wheat is already in the reference to the constellation of Taurus, where the Pleiades are enclosed.

In fact, already in the Vedic hymns, Taurus was venerated as a God, to which is linked the symbolic of fertility complex: horn, sky, water, lightning and rain.

A mythology callback from all the ancient men of all times, not only to build the thick plot of religious traditions, but also for the success of the agricultural season.

In Greek-Roman mythology, Taurus was placed in the sky as he had carried Europe to Crete; Europe itself is the feminine principle fertilized by the male, that is the Moon abducted by Solar Taurus.

This brief final survey to emphasize, once again, the strong link between the mythological tales and the power of Nature, which of course is still relate the Wheat, which brings us the Stone and the Earth, and the Light, which recalls the Sky and the Pleiades.

References

Aveni, A.: Possible astronomical orientations in ancient Mesoamerica. In: Aveni, A. (ed.) Archaeoastronom in Pre-Columbian America, pp. 163–190. Texas (1975)

Aveni, A.: Astronomy in ancient Mesoamerica. In: Krupp, E.C. (ed.) Search of Ancient Astronomies, pp. 165–202 (1977)

Aveni, A.: Skywatchers of ancient Mexico. Austin (1980)

Cattabiani, A.: Florario. Miti, leggende e simboli di fiori e piante. Milano (2010)

Coe, M.: Native astronomy in Mesoamerica. In: Aveni, A. (ed.) Archaeoastronom in Pre-Columbian America, pp. 3–31. Texas (1975)

Droy, V.: La prima mappa stellare europea. Una misteriosa meccanica celeste incifrata nel disco di Nebra, in Archeomisteri 17, pp. 50–58 (2004)

Esiodo—Le opere e i giorni, Milano (1979)

Gaspani, A.: Il Disco di Nebra a 10 anni dalla scoperta, in Coelum XII, nr. 121, pp. 32–37 (2008)

Gaspani, A.: Il disco di Nebra, in I Sentieri dei Celti, Anno III, nr. 5, pp. 32–39 (2009)

Mulas, A., Sanna, M.: In terra come in cielo. I nuraghi e le Pleiadi, Condaghes (2013)
Plinio il Vecchio—Naturalis Historia, libro II, Torino (2010)
Plinio il Vecchio—Naturalis Historia, libro XVIII, Torino (2010)
Ridpath, I.: Mitologia delle costellazioni, Padova (1994)
Virgilio—Georgiche, Milano (2007)

The Figure of the Astrologer in Ancient India: A Practice Verging on the Sacred

Annamaria Dallaporta and Lucio Marcato

Abstract The chapter II of the *Brihatsamhita*, an astrology and divination treatise of the 6th century AD, illustrates the figure and the challenging 'profession' of the astrologer. Backlit a clash arises since the astrologer operated in a context of sacredness considered from Brahmins their exclusive field. That complicated the relations between the two categories particularly at the time of remuneration. According to the *Brihatsamhita*, the astrologer had to be handsome, intelligent, of good birth, well versed in the subject and have the necessary competence. His duty was to identify and interpret the heavenly indications to gain possible profits. For this reason, a king who disdained the services of the astrologer would inevitably be dogged by misfortune. Aside from the risks and misunderstanding implied in the profession, being a good astrologer would bring the maximum reward possible: certainly, at his death he would attain salvation.

1 Introduction

Sacrificial acts, carried out with elaborate rites and codified in every detail, played a very important religious role in India from the Vedic Period.

Sacrifices had to be made on favourable days and at favourable times, purposely 'calculated' by the astrologers on an astronomical basis; failure to observe these recommendations would lead to terrible retribution also for the person officiating.[1] In addition to the sacrifice itself, by extension all other acts of a certain importance depended on the indications of the astrologer who, due to the recognized importance of his services, became a prestigious and (almost) indispensable figure. The

[1]For example, the painter had to begin his painting on a propitious lunar day, under the protection of a benevolent asterism, with the moon in the Citra *nakshatra* (asterism) (*Citrasutra*, chap. 40, 11), a circumstance that was indicated to him by the astrologer.

A. Dallaporta (✉) · L. Marcato
SIA, ISMEO, Indian Archaeological Society, New Delhi, India
e-mail: sarasvati@tin.it

A. Orlando (ed.), *The Light, The Stones and The Sacred*,
Astrophysics and Space Science Proceedings 48,
DOI 10.1007/978-3-319-54487-8_12

role of the astrologer broadened to include guidance, both direct and indirect, on how to behave and on all aspects of life, and formulation of the horoscope became the maximum expression of this activity.

A treatise on the art of horoscopes, *Yavanajataka*, has been dated to the 2nd century AD; its origins are professedly Greek, but Indian predictive astrology, based on the lunar conjunctions, was already practiced in the first millennium BC.

Astronomy in India was generally confined to meeting the needs of astrology more or less up to the 5th century AD when, due to a change in cultural conditions, it regained its autonomy and scientific authority, stimulated by the circulation of Greek, Arabic and Chinese astronomical knowledge.

2 The Characteristics and the Social Position of the Astrologer

Chapter II of the *Brihatsamhita* (*The great compilation*), an astrology and divination treatise of the 6th century AD, is entirely dedicated to the figure of the astrologer, beginning with his physical appearance.[2] We learn that he had to have noble origins, i.e. not belong to the lowest caste of the *shudra*, have a pleasant appearance, present himself amiably, be eloquent, sincere, not harbour envy and jealousy and… have a full voice!

Originality and imagination were also part of the identikit and at the cultural level, an in-depth knowledge of expiation rites and magic was essential, as this would help him to deal with any type of problem caused by the interference of spirits and divinities.

The *Citrasutra*, a 6th century compendium of rules for painting, states that "ministers, Brahmins, astrologers and *purohita* (court priest) must be represented as intelligent persons, with the fair complexion of the lotus flower, a solemn bearing and well-developed body", all characteristics that correspond to those of the second (in order of importance) of the five types into which 'great men' are classified, the first being reserved for kings (III, 42, 2–5; III, 36, 2–3).

In short, he had to be handsome,[3] intelligent, of good birth, have a good character, be well-versed in the subject and have the necessary competence. Said by Varahamihira, the astronomer-astrologer author of the *Brihatsamhita*, it sounds fairly self-referential, but this was the practice.

[2]The treatise is attributed to the astronomer-astrologer Varahamihira and consists of three parts dedicated to astronomy, horoscopes and astrology. In chapter II the physical description of the astrologer is a kind of 'necessary act' which reproduced a sort of formula used also in other treatises (Kramrisch 1999).

[3]The pleasant figure and manners betokened noble-mindedness and wisdom according to the beautiful-good-capable identity as opposed to the ugly-bad-incapable identity, known also in the West.

In the treatise it is not explicitly stated that the astrologer in question is the one at the service of the king and the affluent classes, an astrologer with a capital A, whose requisites were specifically indicated and illustrated to avoid the risk of confusion with the self-styled astrologer ('astrologer'). The latter was more of a magician and diviner than anything else, without any credentials (which nobody would ever have demanded) to exercise his impromptu activity among the masses and scrape a living by begging if necessary, which the 'astrologers' and other underprivileged categories were officially allowed to do (Pieruccini 1988).

In an important sacred text that dates to the 1st century AD, entitled *Manusmriti*, "The laws of Manu",[4] the 'astrologer' is included in the group of people who were not admitted to the honours and the sacrificial meal of the rites dedicated to the deceased (III, 162) which equated to excluding him from that type of ceremony, together with a multitude of other characters who were considered unworthy for various reasons. In another passage (VI, 50) the practice of astrology in order to earn a living is declared to be decidedly unseemly, almost contaminating, for those who, having renounced the world, dedicated their lives to study of the Veda (the so-called *sannyasin*). This only increased the dislike perhaps already felt by the priestly caste ("The laws of Manu" had been written for them) vis-à-vis the 'official' astrologers whose indications they were required to follow meticulously.

In the *Mahabharata* (V, 35), which narrates events at the beginning of the first millennium BC, the 'astrologer' was considered on a par with a person who kills a priest and, as we know, it was difficult to imagine anything worse than this. It sounds a rather exaggerated judgment but it reflects the disdain with which, from the top of the social scale, the street 'astrologers' were looked upon, like jugglers, acrobats, gamblers, crawlers and other individuals of doubtful morality.

We will not analyze here the figure of the 'astrologers', charlatans occasionally mentioned in the *Brihatsamhita* solely to highlight the fact that the only true astrologers are refined, educated, competent people who know the 'science' and the *Veda*. This is why they are sought after by those who are sagacious and regard them highly, and can obviously afford the cost of their services.

The *Brihatsamhita* also contains some traces of the diffidence of the Brahmins vis-à-vis the court astrologers and in general those at the service of the rich and powerful, due probably to the fact that their activity was often closely related to unknown and fearsome forces and even to the world of magic. In addition, the astrologers were disliked because they shared with the priests (but without the support of their authority) an indirect 'control' over governors and administrators and because they had a type of power (already mentioned) consisting in the fact that often it was they who decided on the conditions in which the priests themselves operated. It should be added that, in all likelihood, underlying the practice of

[4]The treatise, put together perhaps between the 1st and 3rd century AD, "constitutes the first systematization of the Hindu law and is the first of the *Dharmashastra*, the Brahminic treatises on the law. (…) It contains the social, ethical and moral precepts that must govern the lives of individuals and instructions for carrying out rites and ceremonies" (Dallapiccola 2005); (see Doniger 1996).

astrology was the intention to meet the need to conform every action to the celestial order. The astrologer therefore operated in a general context that can be considered sacred although, upon closer examination, this context took on a magic-propitiatory character. The Brahmins' perception of an interference in their exclusive religious sphere undoubtedly complicated the relations between the two categories.

In the *Brihatsamhita* it is reiterated several times that, in order for given predictions to have a successful outcome, certain rites must necessarily be officiated by Brahmins to whom plentiful and precious offerings must be made. Astrologers and Brahmins therefore in effect needed each other and this complementarity certainly did not justify a conflict, had it not been for one delicate question: to what extent were the two functions comparable? Formally it was not a problem of caste (although in reality everything was connected with caste) but of role.

Every opportunity was used to re-establish the 'right' distances, in favour of the priests, naturally, thus reaffirming beyond all possible doubt a supremacy that was unquestionable.

A prescription of the *Arthashastra* (V, 3, 13), a famous treatise of the 4th century AD on good government,[5] established that the court astrologer was paid one thousand *pana*, like the staff who were under the direct authority of the *purohita*, positioned at the highest level of the state administration, on a par with the Crown Prince and the king's personal advisers.

Incidentally, the *purohita* received forty-eight thousand *pana*. This is cruel proof of the hierarchy, but of course, everyone furthered their own cause as best as they could.

Again, in the *Arthashastra* the description of the plan of the ideal city is significant: the residence of the *purohita* was to the north-east immediately at the back of the royal palace (II, 4, 8) and that of the priests was to the north (II, 4, 14) whereas there was no indication of the residence of the astrologers.

According to the *Brihatsamhita* the house of the court astrologers should have the same characteristics as those of the priests and royal doctors (LIII, 10): a desire for levelling that had to be expressed, as a sort of official duty or maybe a reference to a social position, of which only the memory remained. This is testified also in the stories of the *Mahabharata*, several centuries earlier, which tell us that the court astrologer was one of the seven highest-ranking persons in the state, together with the *purohita*, the doctor, the governor of the citadel, the military commander, the commander of the guards and the supreme judge (II, 5).

After the beginning of the first millennium BC, with the passing of time, the esteem in which astrologers were held appears to have decreased considerably.

However, in chapter X of the *Samarangana Sutradhara* (a treatise on architecture at the beginning of the 11th century AD) the situation appears to be once again reversed: the residence of the astrologers and priests was relocated to the north, in the same district, whereas the residence of the diviners and fortune tellers (the plebeian 'astrologers') was towards the western edge of the city together with

[5]Treatise on politics and good government written in the Mauryan Age (4th–2nd century BC).

the residences of the elephant keepers, the artisans and the artists. Perhaps this should be regarded as a slavish transcription from a much older text (of the same treatise) or perhaps it represents a real change in the social position of the astrologer, who went back to enjoying his initial status, following the first Muslim conquests of the 11th century.

Given the scanty evidence, these are only hypotheses that will have to be adequately verified if necessary.

According to the sources, however, the Brahmins do not appear to have practiced astrology, otherwise there would be no justification for the irritation they felt with the astrologers, who reciprocated as soon as possible even though they were aware of their inferiority. As indirect proof of this assumption a verse of the *Brihatsamhita* reads: "The astrologer who knows the Veda and the meaning of the science deserves to be respected like a Brahmin and be one of the first to be served in the *shraddha* ceremony (for the ancestors) because his presence purifies the sacrificial meal" (II, 14).[6] To reinforce the concept, the next verse insists on the respect due to the astrologer 'as if he were a priest' and if, despite the counsel, anyone was foolish enough to doubt his word and deride his 'science', a very explicit curse would ensure torments of all kinds in the darkness of the underworld (II, 18).

In our opinion the meaning of the calls for respect (astrologers) and supremacy (priests) had little to do with a question of principle: at least originally they probably concerned, very prosaically, the (demanded) right of the astrologers to benefit, to some extent, from the sacrificial offerings, which were always substantial.

2.1 Knowledge of Astronomy

The *Brihatsamhita*, after indicating the physical characteristics required of a person beginning to learn the 'profession' of astrologer, reviews (again in chap. II) without any attempt at a systematic approach and very succinctly, the astronomical and astrological competences he should possess (discussed in further detail and developed in the subsequent chapters of the treatise).

The fundamental requirement in terms of astronomy for an astrologer was knowledge of the five 'revealed' astronomical systems (*siddhanta*)[7] called: *Surya*, *Romaka*, *Pulisha*, *Pitamaha* and *Vasishtha*, in order of importance.

These theories referred to five treatises (5th–6th century) condensed by Varahamihira in his *Panchasiddhantika* which is considered, in the history of Indian scientific literature, the second treatise of astronomy, in chronological order, after

[6]The 1884 translation from the Sanskrit by N.C. Iyer has been tidied up by the authors.

[7]Astronomy-astrology constituted one of the six *Vedanga*, treatises considered complementary to the *Veda* and, like them, 'revealed'.

that of Aryabhata[8] entitled *Aryasiddhanta*, written shortly before (Boccali et al. 2000). Unfortunately we know almost nothing about these five treatises: a summary contained in the *Panchasiddhantika* refers to the theories of Greek astronomy quoted in the *Romakasiddhanta* and only traces of the *Suryasiddhanta* have been preserved in the treatise of the same name which has survived up to modern times.

Prabodhchandra Sengupta in the *Introduction* to an English edition of the *Suryasiddhanta* of 1935 identified three main phases of formation of the text: the first coinciding with a work already existing at the time of Varahamihira, perhaps of Babylonian origin; the second which Varahamihira drew from the previous one, including among other things the theory of the epicycle; the latter, extensively modified and updated over the centuries several times and by several different hands, is the work known today (Gangooly 2000).

In addition to the contents of the five treatises above, the astrologer had to be familiar with and know how to calculate the subdivisions of time, correctly define the position and course of the sun and the planets by direct observation and the use of suitable instruments and apply the appropriate correction to conversion of the heliocentric latitude into the geocentric latitude and vice versa. He had to be able to predict eclipses, calculate their duration and know their characteristics; using the most appropriate calculations, he had to be able to predict: the conjunctions of the moon with the planets, of the planets with one another and of the planets with the stars; the times when the zodiacal constellations rose in the sky. A few other topics completed the list which in reality did not amount to a great deal, but nevertheless this set of astronomical knowledge was vital to the professional credibility of the astrologer since "his predictions certainly cannot fail if he has a good knowledge of astronomy" (Iyer 1987).

There was probably a desire to spread this concept outside the world of the astrologers, in order to provide a scientific guarantee of the very subjective and questionable activity of the astrologer (the 'science'). The exchange of names in some passages of the text (e.g.: XXI, 3; 4 and following) where the astrologer is called 'astronomer' should perhaps be interpreted in this sense: it could be an error of translation but in our opinion it could have been perfectly intentional since self-esteem was to be stimulated at every possible opportunity (resorting to ploys if necessary) also at subliminal level. This was probably the purpose of the exaggerated adulation, in the *Brihatsamhita*, expressed by an apprentice called Vishnugupta[9]: "Only a great sage is able to imagine crossing the vast ocean of knowledge of the astrologer" (Iyer 1987).

[8]Aryabhata (5th century AD) was the scientist who gave a new and decisive impulse to Indian mathematics and astronomy. He founded two schools of astronomy and wrote some fundamental works, translated into Arabic in the 8th century.

[9]'Professional' treatises like the *Brhatsamhita* had a narrative structure in which a great master, if not the same divinity who presided over that art, explained the concepts and the fundamental rules of the subject, replying to the questions of one or more disciples.

In reality the figures of the astronomer and the astrologer could coincide in the same person as was normal also in the West at least until the 17th century.

However, all this appears to indirectly point to the conclusion that a knowledge of astronomy in itself was considered in ancient India fruitless unless a 'useful' application was found, as in astrology. It should be remembered that for a Hindu, 'utility', in all its possible meanings, constituted one of the main existential values.

2.2 Astrological Knowledge

While the knowledge and measurement of time was considered to belong to the sphere of astronomy, the ability to give space a sense, by identification and definition of the directions (orientation), pertained to astrology.

For the formulation of the horoscopes, the astrologer had to know in particular the planets, their aspect, their position, their relationship with the various parts of the body and with human destiny. He had to be able to predict the future, the duration of life and the cause of death of a person. For this, he needed great skill to consult a handbook (called *Ashtakavarga*) which from the combination of some elements (like the date, time of birth and position of the planets relative to the signs of the zodiac at the moment of the event) enabled him to formulate the required predictions.

The task of the astrologer was also to indicate the auspicious periods for ceremonies, particularly for marriage; predict storms and earthquakes, the beginning and end of the monsoon, the quality and quantity of the harvest.

Via the interpretation of particular signs of the ground and the environment, of natural phenomena and dreams, the royal astrologer had to establish the places in which to deploy the army or the most suitable time to begin a battle which was undertaken only with the presage of victory. In this regard, to ascertain the fortunes of the king in a battle, it was customary for the first minister, the priest and the astrologer to sleep in the royal temple; the astrologer would then interpret the dreams of each one of them to find a common meaning.

The prediction of future phenomena was also based on the interpretation of the flight of certain birds, the behaviour of antelopes, dogs and various other animals.

All this required a huge number of different competences which certainly could not be expected of one single 'professional'. The quantity of observations and predictions to be formulated and the abilities called for required, in all likelihood, at least another four expert collaborators, adequately remunerated (Iyer 1987).

Regardless of the actual need for assistants, we believe that they provided above all a scapegoat for any predictions that did not come true, with detrimental consequences: the astrologer would save his own head and his position since all failures would be easily attributable to an imaginary shortcoming of his assistant. It was probably not as easy to lay the blame on the priest for inaccuracies in celebration of the rites that ensured the effectiveness of the predictions, but it is likely that all strategies became feasible in order to get out alive and in a more or less dignified manner. The position of royal astrologer was certainly much sought-after but the so-called 'risks of the trade' were very serious and therefore had to be neutralized in every possible way.

It is reasonable to assume that when it was necessary or simply expedient, the success of the predictions also depended on subterfuge and all sorts of ruses purposely orchestrated by the astrologer, but there is no trace of this in the literature. In India, faith in astrology is still deep-rooted and widespread and is even studied as a subject in some universities. Scepticism and disbelief, like ours, would therefore not elicit any type of support or interest.

The last part of chapter II was addressed indirectly, but with a certain insistence, to the generic wise and enlightened governor to convince him, if necessary, to make permanent use of an astrologer. This was done for his own good because a king who disdained the services of the astrologer (evidently not everyone placed their faith in his 'science') would inevitably be dogged by misfortune (II, 7) and would be destined in life and in the government of his country to flounder disorientated in complete darkness (II, 9). His refusal simply meant that he did not desire for himself success, fame, health and happiness, and for his people a serene and prosperous existence (II, 11–12), values which were priceless. Even only one astrologer (if not aided by the four assistants) could render his master an enormous service which, in quantitative terms, not even one thousand elephants or four thousand horses could ever offer (II, 22).

There was no reason why there should not be at least one astrologer at court; in fact, neither the parents, brothers or friends of the king could desire his wellbeing like a 'true' astrologer could (II, 24). 'True' meant not superficial, expert, well-versed and therefore reliable: anyone who passed himself off as an astrologer without adequate study and preparation (the self-styled astrologer as above) should simply be considered a rogue and a social disaster (II, 17).

In short, between the lines the king was given to understand that if he did not make use of an astrologer, it would be a risky unforgivable act of foolishness. Obviously the interest of the king was matched by a much more concrete interest of the astrologer, as is candidly shown, for example, in the chapter on the periodic lustral bath: at the end of the ceremony known as *pushya homa* "the king will gratify the astrologer and the priest by giving them large sums of money" (XLVIII, 80).

Aside from the risks and misunderstandings of the 'profession', being a good astrologer would bring the maximum reward possible (and here we have the prediction of all predictions, highly consolatory): at his death he would go to the Paradise of Brahma, attaining salvation (II, 13).[10]

All the topics briefly mentioned so far are then taken up again in greater or lesser detail, as we have said, in the other 104 chapters of the treatise (there are 106 chapters in all) which amounts to a list of the significant situations of different kinds with the relative interpretations, which follow one another in the text in a fairly disjointed manner. Nine chapters are dedicated directly to the planets with

[10]'Salvation' here means release from the cycle of rebirths.

indications of this type: "A single spot on the solar disc can result in famine for the population, two or more decree the sudden death of the sovereign. According to their colour, the spots will also be a cause of great suffering: for the Brahmins if white; for the *kshatria* (warriors) if red; for the *vaishya* (traders) if yellow; for the *shudra* (manual workers) if dark" (III, 19). Or: "If the disc of Jupiter/Brihashpati appears large, bright, jasmine white or water-lily white colour, well visible, not in conjunction with other planets, humanity can live happily" (VIII, 53).

As regards the planets, situations that could lead to disasters and calamities and others leading to prosperity and fortune are variously distributed. Predictions caused by Mercury/Budha, Saturn/Shani, Rahu and Ketu[11] were generally negative, and often there was nothing good to be expected from Venus/Shukra.

The chapter V, dedicated to Rahu, is the longest of those concerning the planets and at the beginning contains some surprisingly 'modern' verses in which scientific truth is made to prevail over the myth.

The legend of the monstrous head of Rahu roaming the sky, ready to devour the sun and the moon thus causing the eclipses, is introduced by an anonymous 'people say that'. A few lines later, however, we find a revolutionary affirmation (for the context in which it appears): the truth lies not in the 'people say that' but in the fact that a lunar eclipse occurs when the earth is positioned between the sun and the moon, and the solar eclipse when the moon is positioned between the earth and the sun (V, 8). In other words, eclipses are caused not by the demon Rahu but by the interpositions studied and demonstrated by able astronomers (V, 13).

The reassuring traditional truth was abruptly disowned and supplanted by a new scientific truth, totally alien to the myth on which the reason for existence was based.

The first 18 verses of the chapter V, in which science is made to prevail over the traditional knowledge, are like a foreign body, detached from the logic not only of the other 80 verses of the chapter but of all the chapters that illustrate the 'science' of the astrologer. It is therefore probably an addition included almost forcibly, after the first drafting of the treatise, and the high number of verses also point to this hypothesis.

Chapter V is the longest of those dedicated to the planets, and in terms of the entire treatise, the highest number of verses is contained in chapter LIII "On building of the house" (125) and in chapter LIV "On underground currents" (125) followed closely by chapter LXVIII "On human features" (117). The latter opens with a declaration: "The expert astrologer, by examining the body, the complexion, the voice, the vigour, the articulations, the brilliance, the colour and

[11]Rahu and Ketu were considered planets of past aeons: their position coincided with the ascending and descending nodes of the lunar orbit. They were personified, the former by the demon of the eclipses represented by the sole monstrous head, and the latter by the demon of meteors and comets, with the body of a snake. Eclipses, meteors and comets were generally perceived as phenomena of ill omen.

the form of the face, and the height, weight, character and gait of a person will be able to predict his destiny" (LXVIII, 1), then discussing each item in minute detail (unintentionally often humorously) and forming a sort of corollary consisting of the surprising chapter LII "On boils" (10).

The astrologer also had to be an able interpreter of boils since, combining the caste and sex of the person, the part of the body (none excluded!) in which the boil was located and the color of the growth, it was possible to establish how and to what extent it would influence the life of the person in the immediate future. It should be noted that boils and their effects were grouped together with black-heads, warts and the like (LII, 10).

3 Conclusions

One might say, like the apprentice *Vishnugupta*, that there was no limit to the knowledge of an astrologer. This appears to be confirmed by other chapters on all sorts of different topics such as: "On the price of consumable goods" (XLI), "On the sword" (L), "On gardening" (LV), "On representations in the temples" (LVIII), "On perfume blends" (LXXVII), etc.

More realistically, however, it seems that, in addition to the distinction between "official" and "street" astrologers, some were more able in certain areas than others, and each one limited himself to uniform and more or less defined fields.

It is not particularly surprising that an astronomer, on the basis of his observations and a practice that was easily acquired, was able to produce horoscopes and predictions[12]: it would be a little more surprising to know that he was a haruspex or geomancer. In the same way we can imagine an astrologer who concentrated his activity mainly on indications concerning the construction of dwellings and temples: according to Stella Kramrisch, architecture could be considered as astrology applied to constructions (Kramrisch 1999). Astrologers could be particularly versed in the problems connected with the army and warfare, or matrimonial situations, others in horoscopes and so on, since the possible fields of activity, like those mentioned above, presupposed extensive knowledge and experience.

With the advent of Muslim culture and tradition (from about the 11th century) these different sectors would become specializations of independent branches of a complex and very wide-ranging 'science'.

[12]In addition to this, it should be remembered that, right up to the 17th century, the mathematicians-astronomers were called on to teach astrology in the medical faculties of the universities.

References

Boccali, G., Piano, S., Sani, S.: Le Letterature dell'India. UTET, Torino (2000)
Dallapiccola, A.L.: Induismo. Bruno Mondadori, Milano (2005)
Doniger, W.: Le Leggi di Manu. Adelphi, Milano (1996)
Gangooly, P.: The Surya Siddhanta. Motilal Banarsidass, New Delhi (2000)
Iyer, N.C.: The Brhat Samhita of Varaha Mihira. Sri Satguru Publications, Delhi (1987)
Kramrisch, S.: Il Tempio Indù. Milano–Trento, Luni Editrice (1999)
Pieruccini, C.: Dandin—I dieci principi. Paideia Editrice, Brescia (1988)

The Star of the Sibyl: Analysis and History of a Late Medieval Illustrated Prophecy

Giangiacomo Gandolfi

Abstract Misinterpreted and overlooked in its astrological meaning for a long time, the so-called "caricature of the Emperor and the Pope" of the late XVth century is actually a complex stratification of joachimist traditions and at the same time an extraordinary example of how an anonymous and ambiguous engraving imbued with apocalypticism could have left a mark on the history of culture, spreading and adapting across the centuries. I propose here an astronomical and textual analysis, trying to identify the sources of the prophecy and the nature of the star represented on the basis of the earliest german and venetian versions. As a result of the inquiry I trace an evolutionary path strictly connected to the myth of the Tiburtine Sibyl that reveals a complicated and muddy transmission between Bohemia, Venice and Germany and involves personalities such as Johannes Capistranus, Cornelius Gemma and Tycho Brahe. The persistent resurgence of the prophecy at different times and in association to astrology, testifies not only its political role in nationalistic and religious propaganda, but even a symbiotic relationship between cosmologic research and chiliastic concerns at the threshold of the Scientific Revolution.

1 Introduction

The venetian engraving published by Arthur Hind in his celebrated catalogue (Hind 1938), reproducing an alleged carved stone found in Altino in 1495, is a unique and fascinating example of illustrated prophecy of the late Middle Ages and at the same time a seminal effort in connecting the tradition of the Sibyls to the science of the stars. As effectively shown by Ackermann Smoller (2010), such a tradition, unrelentingly revisiting the oracles of the classical prophetesses of antiquity, became at the threshold of Renaissance progressively entangled with astral divination and

G. Gandolfi (✉)
Planetario e Museo Astronomico di Roma, Rome, Italy
e-mail: giangiacomo.gandolfi@gmail.com

A. Orlando (ed.), *The Light, The Stones and The Sacred*,
Astrophysics and Space Science Proceedings 48,
DOI 10.1007/978-3-319-54487-8_13

provided a basis for a curious form of "astrological natural theology, sacralizing Science as well as Nature" (Ackermann Smoller 2010).

I propose here an updated analysis of the image and of his german sources, trying to emphasize the two most relevant aspects of the prophecy, namely its evolution and adaptation to different historical scenarios and its connection with the technicalities of astrology. Following the interpretive path of previous scholars as Smahel (1994) and Vassilieva-Codognet (2012) and developing a specific astrological model I will also suggest an area of origin and two plausible dates for the earliest commented versions, if not for the prophecy itself.

2 Astrological Sibyls

The Graeco-roman tradition of the Sibyls, with variable lists containing from 2 to 30 prophetesses,[1] was transmitted to the medieval culture through the mediation of Augustine and of the Fathers of the Church. The role devised by Christian authors for these enigmatic women was propagandistic: sibyls were the counterparts of biblical prophets and attested the redeeming power of Christ since the most remote Antiquity, domesticating and assimilating the pagan world. During the Renaissance the sibylline presence in art and literature reached its peak, but its relevance was gradually undermined by humanistic philology, ever more effective in revealing the fallacies of chronologists comparing the Bible to Classical works. The increasing blending of astrology and prophecy, on the other hand, transformed some of these figures in a sort of feminine prognosticators and established a new image of the prophetesses as women looking intently at the sky, a real "narrative of the star of the Sibyl". Such a narrative was declined in two different settings, or in a mixture of them: apocalyptic prophecy (the model are the Oracula Sibyllina (Geffcken 1902), a medley of Jewish apocripha of the II-I century BC mixed with much later Christian and Gnostic fragments) and edificatory literature (aimed at confirming and exalting Christ's Coming).

A perfect example is the Sibylla Chimica (or Cimmeria) in the early *Prophetie XII sibillarum de incarnatione Christi* (Hélin 1936) and then in the famous catalogue of Barbieri (1481), which is connected to a text of the Arabic astrologer Albumasar foretelling the Incarnation in the first decan of the constellation Virgo,[2] while astrological elements abound in the german *Sibyllen Weissagung* (Neske 1985) and in the famous apocalyptic *Pronosticatio* of Lichtenberger[3] where a Sibyl guarantees with other prophets the veracity of astral predictions.

---

[1]Varro codified 10 Sibyls. For a general treatment of the Sibyls see H.W. Parke, *Sibyls and Sibylline Prophecies in Classical Antiquity*, NY, Routledge (1988).

[2]On this tradition, connected to the doctrine of the Great Conjunctions, it is particularly useful Pompeo Faracovi O., *Gli Oroscopi di Cristo*, Venezia, Marsilio (1999).

[3]For a recent study see G. Petrella, *La* Pronosticatio *di Johannes Lichtenberger: un testo profetico nell'Italia del Rinascimento*, Udine, Forum (2010).

The favorite figure for this kind of contamination is the Tiburtine Sibyl Albunea. Her connection with the stars is attested since roman times, when Tibullus associated the prophetess with a malevolent comet,[4] but the tradition was reinforced in late antiquity, when since the VIth century a legend about a seeress showing to Augustus a virgin with child in the sky above the Capitolium began to circulate, with the woman becoming Albunea and the vision transforming in a kind of parhelion encircled by stars during the Middle Ages.[5] Furthermore, a famous apocalyptic prophecy of eastern origin—the so-called "dream of the nine suns"[6] (IVth century), was attributed to her in latin translations, again pointing to celestial inspirations.

Particularly important for the analysis of the Altino Prophecy is the somehow parallel tradition of the Queen of Saba, which was often considered a thirteenth Sibyl,[7] well versed in the science of stars and famous for having predicted to King Solomon the Passion of Christ, as narrated in the famous *Legenda Aurea* (Lazar 1960).

3 The Altino Prophecy

The three venetian engravings containing the prophecy, the most discussed by art scholars,[8] are remarkably similar, but just two of them contain the date of the presumed translation of the carved stone to Venice: the fateful year 1495, characterized by a bloody series of Italian wars. The only exemplar lacking the date has been for a long time considered on the basis of the style a print produced in Venice around 1470 and interpreted as a caricature of the meeting of the pope Paul II and the emperor Frederick III (Hind 1938). While it is true that the engravings dated 1495 are not the original version of the scene, we are now reasonably certain that the undated exemplar is contemporaneous to the other two and that many elements of the composition support the prophetic nature of the image, and the absence of any satirical intent (Vassilieva-Codognet 2012, pp. 199–206). On the other hand,

---

[4]Tibullus, *Elegiae*, II, 5: "*[…] quaeque Aniena sacras Tiburs per flumina sortes portarat sicco pertuleratque sinu/haec fore dixerunt belli mala signa cometen, multus ut in terras deplueretque lapis*".

[5]An excellent analysis of this narrative is the anonymous *El Enigma de la Sibila*, pp. 28–45 https://sites.google.com/site/omnedecus/Home/art/el-enigma-de-la-sibila.

[6]Holdenried A., *The Sibyl and her Scribes*, Aldershot, Ashgate (2007), a complete census of all the versions of this prophecy in manuscript and in print.

[7]El Enigma de la Sibila, *op. cit.*, pp. 69–79.

[8]The third version is preserved in the National Gallery of Washington. For a discussion of all the exemplars see J.A. Levenson, K. Oberhuber et al. (eds.), *Early Italian Engravings from the National Gallery of Art*, Wahington (1973), n. 69 and P.O. Kristeller, "Der venezianische Kupferstich im XV. Jahrhundert", *Mitteilungen der Gesellschaft für vervielfältigende Kunst* (1907), pp. 1–16.

Fig. 1 The Altino Prophecy (London—Brit. Museum (Hind E. III. 6); London—Brit. Museum (Hind E.III.7); Washington—National Gallery of Art)

the source of the prophecy has been traced in a series of pamphlets printed in south Germany in the 70s and in a drawing found in a famous venetian collection of prophecies,[9] leaving open the hypothesis of a connection with the meeting.

In order to reorder the sequence of different versions of the prophecy and to collocate them in the right historical and cultural context, a careful analysis of their elements has to be carried out, beginning with the bare image and only later dealing with the texts, that could have been added to the scene by an anonymous commentator (Fig. 1).

3.1 Analysis

The centre of attention of the composition is undoubtedly the ambiguous relationship of the pope (identified by the *triregnum*) and the emperor (identified by the crown), both positioned on the top of the main mast of a ship. Are they wrestling or helping reciprocally in a situation of unstable equilibrium? What is certain is that they lean respectively on a wheel and on a lion, while the emperor shows an undeniable attitude of submission. The "kaiser" apparently holds the broken spindle of the wheel sustaining the pope, but analyzing the earlier german images it is easy to notice a defect of transmission and to conclude it is a broken scepter instead. From his neck hangs a sack, clearly indicating avarice. On the contrary the pope holds a scale and the coat of arms of the Valois dynasty and is menaced by a snake encircling his head. On the left side a bright star loom over the scene hitting with his

[9] The folio 10v of Cod.Lat. III, 177. On this collection see M. Reeves, *The Influence of Prophecy in the Later Middle Ages*, Oxford (1969), pp. 343–346.

rays the lion, while up on the right stands the double eagle of the Sacred Roman Empire and of the Habsburgs. The ship, which is probably the *navicula petri* of the Church and for transitivity the symbol of Christianity, is heading towards a dead tree, where the French king shield decorated with lilies is hanging.

So far we have ignored the textual comments, that in many cases confirm the evidence: the two figures are labeled *papa* and *imperator* (in the german leaflets *Fredericus Imperator*), the coat of arms *rex francie*, the star almost always *cometa* and the double eagle sometimes *imperium*. Equally reasonable appear the *civitates imperatorie*, *duces Austrie*, *electores imperii*, *rex Hungarie*, *rex Poloniae*, *Buaguera* (Bavary) *duces Budrium* (Buttrium), *rex Dacie*, *dux Sacsonie*, *rex S(acsonie)-A(nhalt)* and *dux Schotie*, all inscribed on the ship, while the cities on the ladder onboard (*Ragosa*, *Venetia*, *Bosina*, *Sicilia*) are more enigmatic. Other labels, conversely, are surprising and revealing: the snake is associated to the *dux Mediolani*, the lion touched by the comet to the *dux Burgundiae*, the broken scepter to the *rex Bohemiae* and the scale to *Roma* (with *SPQR* on the plates). Furthermore the dead tree has a *Hierusalem* written on the trunk and the papal wheel is accompanied by a mysterious *patriarchae rome*.

Finally, four longer texts are distributed across the image, three of them in both german and venetian exemplars: the fourth, present only in the Italian engravings, associates it to a prophecy of the Tiburtine Sibyl. The undated exemplar reads: "*Prophecia dela Sibilla Tiburtina trovata in una granda pietra in la città di Altini in uno loco ruinato, fata inanti l'avenimento al nro sigr Iesus Christus ani 19 così como qua retrata e scolpita e fo traslata a Venecia*", referring to an imaginary roman art work of 19 BC.[10]

Under the star two texts are found in all versions: "*Sub Saturno in domo infirmitatis*" and "*Surculus est illa de qua sibylla prophetavit hoc presagioque beavit*", while the double eagle is encircled by the phrase: "*Sum quod eram nec eram quod sum iam dicor utrumque*", an ancient theological riddle often referred to the nature of the Virgin Mary, here probably intended as a label for a French king becoming Imperator and as such holder of both qualifications (Vassilieva-Codognet 2012, p. 201).

3.2 Interpretation

We have already noticed that far from being a satirical portrait of two actual historical figure, the Altino image is an illustrated prophecy. Even if creators and commentators of the figure may have thought of Paul II and Frederick III meeting in Rome under the bad auspices of the 1468 comet while a crusade against the Turks was being hoped for (giving a supplementary meaning to the lion under the emperor's foot), this is clearly just a possible outcome of the prophecy, which instead has to be more generally

[10]The others two versions, as explained before, cite the year 1495.

intended as a symbolical system deeply embedded in the language and in the tradition of vaticinia. Such a context is intrinsically polysemous, and allows multiple choices for the historical figures that embody the symbols.

This is exactly the problem one has to face in trying to interpret historically the prophecy: the success and the continuous resurfacing of the image across the centuries mean that several readings are possible, making the analysis complicated and never conclusive. That means that the right question does not directly concern the figures of the Europe of XVth century that may be portrayed in the engravings, but regards firstly the models and the sources of the prophecy, the cultural contexts that gave strength to the representation.

Following this route Smahel and Vassilieva-Codognet have effectively shown that the Altino scene has to be connected to the so-called "Last Emperor"[11] type of prophecies, referring in particular to the complicated bohemian scenario with his hussite dissidence and to the dramatic parabola of the duchy of Burgundy.

The general pattern of this narrative, with his final voyage to Jerusalem and in certain versions the hanging of his shield on a dead, dry tree, is declined in an extremely ambiguous way: the *insignia* are certainly French, and this points to the pro-Gaulish "Second Charlemagne" variation of the legend (Reeves 1969, pp. 320–331), but the double-eagle and the label "Fredericus"[12] seem to indicate that the intended variation is the pro-Teutonic "Third Frederick" (Reeves 1969, pp. 332–346). Such a confusion is not surprising at all, as the "Last Emperor" prophecy was a battlefield for nationalistic factions that included hybrid narratives with frequent contaminations of different conflicting elements, featuring alternately a good Frederick, a bad Frederick, a good Charles and a French antichrist (Reeves 1969, pp. 333–336).

In later reprises of XVIth century the famous "*Prophetia de ultimis temporibus*" aka the Gamaleon Prophecy[13] is explicitly mentioned in the text around the image, stressing the pro-german version and explaining the round object under the Pope's foot as the Mainz Wheel, symbol of the seat of the future reformed Church in Moguntum.[14]

This intricate scenario is however further complicated by the ambiguous confrontation of Pope and Emperor, which is not typical of the Last Emperor

[11]This tradition stemmed from the *Apocalypse of Pseudo-Methodius*. See P.J. Alexander, "The Medieval Legend of the Last Roman Emperor and Its Messianic Origin", *Journal of the Warburg and Courtauld Institutes*, 41 (1978), pp. 1–15.

[12]To this evidence could be added the enormous success of the prophecy in Germany, attested by the large number of prints produced.

[13]For this prophecy see B. McGinn, *Visions of the End: Apocalyptic Traditions in the Middle Ages*, New York, Columbia University Press (1979).

[14]The version of the Gamaleon Prophecy reported by Wolfgang Lazius in his *Fragmentum vaticinii cuiusdam... Methodi...* (Vienna, 1527) mentions explicitly the "patriarchate" of Mainz with this role.

Fig. 2 Precursors of the Altino Prophecy (Venice—Biblioteca Marciana (Lat.III.177, 1469?); Wien—Graphische Sammlung Albertina; Munchen—Staatsbibliothek (Michel, ca. 1470/1480))

narrative,[15] and by the menacing presence of a bright star, associated to an ill-defined sibylline prophecy.

3.3 *Evolution and Adaptation*

As we have seen the ambiguity of the image allows an easy adaptation of the general outline to different historical contexts and ideological positions with minimal semantic shifts. The characters of the prophecy are "functions" rather than real historical figure, as in the case of the emperor Frederick, that only accidentally is a monarch of the Habsburg family and probably embodies the function of a Last Emperor successor of the legendary Frederick II Hohenstaufen.[16]

Nonetheless, on the basis of the dating of the first german pamphlets and of the Marciana drawing that could be considered the precursors of the Altino representation (Fig. 2), the game of historical correlations may be attempted. Reflecting an earlier original, the prophecy could represent Pope Pius II—Aeneas Silvius Piccolomini (1405–1464)—and the same Frederick III of Habsburg (1415–1493), with Charles VII (1403–1461) or Louis XI (1423–1483) as king of France. The lion in this case should allude to Philip the Good duke of Burgundy (1396–1467) and the *rex Bohemiae* to George of Podebrady (1420–1471), whose pro-utraquist positions

[15]Vassilieva Codognet reads this confrontation as the deposition of the german emperor by the pope in favor of the Second Charlemagne (the raised coat of arms with the French lilies would confirm it), but such an interpretation, although seductive is in contrast with the enthusiastic reception of the image in Germany.

[16]The Last Emperor Prophecy was incredibly boosted by the figure of Frederick II, as effectively shown in M. Reeves, op. cit. (1969), pp. 306–319.

broke the harmony of its scepter with orthodox Christianity. Accordingly, the ship and the confederation of states onboard would fit perfectly Piccolomini's project of an european expedition against the Turks, with Venice as an ally and Raguse and Bosnia as stop-over on the road to the Holy Land.[17]

The same scenario could be valid for 1464–1471, with the new Pope Paul II (1417–1471) and Charles the Bold (1433–1477) instead of Philip the Good, while the broken scepter would be even more compelling for a George of Podebrady officially excommunicated and involved in the lacerating Bohemian Wars. One may also stretch these historical interpretations to the year of death of Charles the Bold (the lion struck by the comet), recalling his alliance with the Emperor sealed by the marriage of his daughter Mary of Burgundy with Maximilian of Habsburg (1459–1519). At the time the pope was Sixtus IV (1414–1484) and the king of Bohemia Matthias Hunyadi (1443–1490), still involved in a long war against the successor of George of Podebrady, Vladislaus II Jagellon (1456–1516).

The reprise of the prophecy in Venice around 1495 changed completely the setting and the ideological subtext of the image, being probably associated to the descent in Italy of Charles VIII (1470–1498), a good candidate for the role of Second Charlemagne. The pope was Alexander VI (1431–1503) and the emperor Maximilian I, but the relevance of the broken bohemian scepter (now in the firm hands of Vladislaus with Hungary and Croatia) was clearly diminished, as the role of Philip the Handsome (1478–1506), the Castilian son of Mary of Burgundy and Maximilian of Habsburg. Neither a crusade seemed more forthcoming, yet the prophecy should still have had a strong grip on minds during the Italian wars when spiritual anxiety was ready to accommodate any incongruence of the image.

The narrative of the star of the Sibyl had an interesting resurgence during the first quarter of the XVIth century with the so-called "Oryetour Sydus" prophecy,[18] in a form similar but not coincident to the Altino one. This anonymous and textual vaticinium of the Tiburtine Sibyl seemed to emerge from the swiss mountains in 1520 and concerned Charles V and a bright star appeared in the North, announcing a universal reign, cruel wars, calamities and the final apocalypse. It was so widespread and successful that it is discussed even in the writings of Cornelius Gemma and Tycho Brahe, both convinced of its connection with the nova in Cassiopeia of 1572.[19]

The original illustrated prophecy resurfaced again in many german pamphlets around 1540 and 1556, associated to a newly found independent source: the

[17]This is the interpretation defended by P.O. Kristeller, *op. cit.* (1907).

[18]On this little studied prophecy see F. Secret, "Cornelius Gemma et la prophètie de la Sibylle Tiburtine", *Revue d'histoire ecclésiastique*, LXIV (1969), pp. 423–431.

[19]For this debate see H. Hakansson, "Tycho the Apocalyptic: History, Prophecy and the meaning of naural phenomena", in *Science in Contact at the Beginning of the Scientific Revolution*, Prague (2004), pp. 211–236.

pseudo-Capistranus Prophecy (Fig. 3).[20] Undoubtedly apocryphal, this text was attributed to a Silesian monk that observed a comet in 1460 that seems to coincide somehow with the saint of the same name, very respected in Germany and Bohemia.[21] The picture appears here curiously transformed, with the pope represented in a feminine version as a Babylonian Whore while the comet becomes tailed and fiery and the comments proliferate, citing at length the mentioned Gamaleon Prophecy. The tone here, as in the reprise by Johann Wolf (1600) that reproduces the old engraving without texts and labels,[22] is explicitly tinged of vehement lutheran accents against the Church of Rome and both the Emperor and the Pope are clearly despised. The modified image is lastly published in 1619–20, with an only textual reprise in 1632, without thorny labels and adapting the vaticinium to the new historical season.[23]

3.4 *Origins*

It is possible to trace the origins of the image focusing on its earliest versions, namely the german pamphlets signed by the engraver Michel[24] and the Marciana drawing associated to the collection of manuscripts by Frater Rusticianus, copied in 1469.[25] Analyzing such versions two elements emerge, indicating strongly a bohemian provenience: the sibylline text and the broken scepter.

We have already considered the latter, recalling the bohemian religious crisis[26] and the excommunication of George of Podebrady followed by the Bohemian wars: the representation of this metaphorical break seems a clear indication of a particular interest in Czech internal affairs. But another essential piece of evidence is also offered by the enigmatic phrase citing in all versions a *surculus*—the young branch of a tree—and a *sibylla*, sometimes called Liburna. This name does not correspond to any known prophetess, but it could be an orthographic variation of Libussa, the famous seeress-queen of ancient Bohemia that predicted the ascent of the city of

[20] This prophecy has been systematically overlooked by scholars. It appears prominently in J. Wolf, *Lectionum Memorabilium et Reconditarum* (1600), pp. 824–831. Some notes are available in O. Bonmann, "Zum Prophetismus des Johannes Kapistran, 1386–1456," *Archiv Für Kulturgeschichte*, 44 (1962), pp. 193–98.

[21] St. John of Capestranus was actually associated to the Halley comet that loomed over the siege of Belgrade, where he was an absolute protagonist in 1456.

[22] Wolf, op. cit., p. 824. The image is evidently a fake *Ur-Bild*.

[23] A useful catalogue of all the relevant images, even of the pseudo-Capistranus Prophecy, may be found in Smahel, op. cit. (1994) pp. 68–69.

[24] Active in the decade 1470/80 as reported by W.L. Schreiber, *Handbuch der Holz- und Metallschnitte des XV. Jahrhunderts*, Leipzig (1926), 4, pp.103–104.

[25] But according to the analysis of B. McGinn, "Circoli gioachimiti veneziani", in *Cristianesimo nella Storia*, 7 (1986), pp. 19–39, the folio containing the prophecy is not consistent with the other Rusticianus's images and may be a later addiction.

[26] Pius II revoked in 1462 the *Compacta*, the covenant with the moderate faction of the Hussites (the utraquists), creating many tensions.

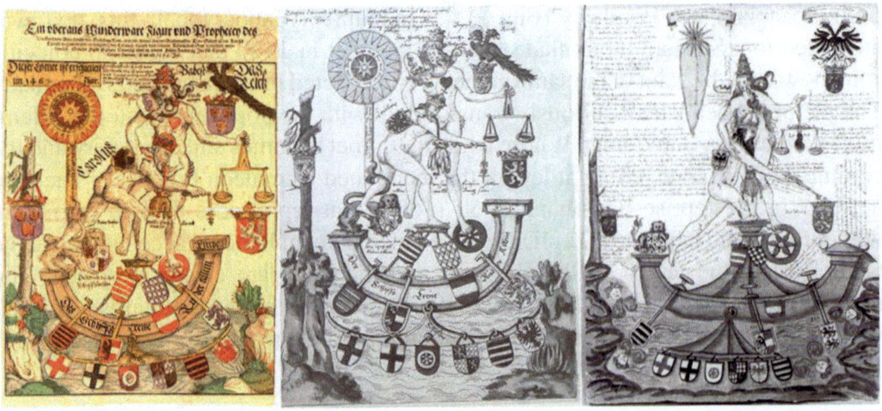

Fig. 3 The Pseudo-Capistranus Prophecy (Munchen—Bayerische Staatsbibliothek (1556); Wolfenbuttel—Sammlung der Herzog-August-Bibliothek (1556); Weissenburg—Stadtarchiv (?))

Prague.[27] In fact the almost identical text "*Surculus est illa, de qua Libussa Sibylla/Sagax prophetavit, hoc presagiumque beavit [...]*" may be found in the old Dalimil Chronicle by Cosmas of Prague in the XII century and later in the "Fontes Rerum Bohemicarum" by Giovanni Marignolli in XIVth century, two bohemian historical chronicles of enormous importance in the forging of a national Czech spirit.[28] Marignolli in particular intended this small poem as a crucial "proroctvi Libusino", parallel to the prophecy of the Tiburtine Sibyl. Libussa foretold the future of Charles IV of Luxembourg king of Bohemia (the surculus, the branch of the Carolingians and the Premyslids through his mother Elisabeth) as Last Emperor in perfect agreement with the standard sequence of the Second Charlemagne prophecy. Furthermore, in Bohemia Libussa's figure has merged with Michalda, the Queen of Saba, the same figure the *Sibyllen Weissagung* considered the thirteenth sibyl and connected again with the Tiburtine Sibyl, a fact that makes the *surculus* resonate with the wood of the cross and the trunk in the Altino image.

But that is not the whole story. The tale of Michalda narrated by the *Sibyllen Weissagung* was translated in the XVth century in Czech in a *Proroctvi Sibyllino* (Verkholantsev 2008) that as the original presents once more many of the elements of the Altino Prophecy: (1) a plot modeled on the dream of the nine suns involving again the Tiburtine Sibyl, (2) a comet that appears in 1471 (or in 1461 in some german version[29]) portending great misfortunes, (3) a Last Emperor reaching

[27]For the myth of Libussa and its political implications see V. Zurek, *L'usage comparé des motifs historiques dans la légitimation monarchique entre les royaumes de France et de Bohême à la fin du Moyen Âge*, Ph.D. Dissertation (2014).

[28]On this two classics of Czech literature see A. Thomas, *Anne's Bohemia: Czech Literature and Society, 1310–1420*, Minneapolis, Un. of Minnesota Press (1998).

[29]Example, in Cod 537 from the Bern Library. See I. Neske, *op. cit.* (1985), pp. 56–58.

Jerusalem and hanging his shield on a dead tree. All evidences pointing to a bohemian influence at least on the first commentators of the image.

4 The Astrological Content of the Prophecy

An overlooked but useful interpretive line of thought may be followed analyzing the astrological content of the prophecy, and in particular the nature of the dominating bright star and the meaning of the allusion to Saturn. Cross-correlating these hidden but precious pieces of celestial information it is possible to arrive at an astral dating that should be carefully compared with the historical scenarios.

4.1 Which Comet?

The historical celestial body represented in the Altino image could be identified—if real—selecting in the contemporary chronicles the comets appeared in a time range compatible with the earliest versions of the prophecy. We present a list composed choosing objects that have some kind of relation with the prophecy, even in later stages of its historical evolution (e.g. a connection with the constellation of Leo and Libra, an association with bohemian and burgundian happenings or with the Capistranus vaticinium, etc.). The list begins with the famous 1456 apparition of Halley's Comet (during the victorious siege of Belgrade, led by Capistranus) and ends with the minor comet of 1477 (year of the death of Charles the Bold in Nancy).

4.2 The Role of Saturn

The historical celestial body represented in the Altino image could be identified—if real—selecting in the contemporary chronicles the comets appeared in a time range compatible with the earliest versions of the prophecy. So far nobody has noticed that the phrase "*Sub Saturno in Domo Infirmitatis*" has a precise technical meaning in astrology, a meaning that refers to an horoscope and to the domification of the theme. The comment indicates something happening when the planet Saturn is in the house of infirmity—the 6th house—and is connected to the comet that stands above this particular fragment of text. Due to the fleeting nature of the position of the houses in a horoscope, this something could not be associated to the first apparition of the object in the sky, because the discovery of a comet, depending often on many observers, could not be exactly pinpointed in time.

Table 1 Cross correlation between candidate comets and revolutions of the year of the discovery, calculated with both the Arabic and Ptolemaic method. In green the Sixth House (Infirmity) and in red the position of Saturn

Year	Sources	Constellation	Revolution of the year (eq.)	Revolution of the year (syz.)
1456 (Halley)	Pingré	Cancer		
1460	Rockenbach, Alsted, Hevelius, Riccioli, Praetorius	? + Ecl ⊙ 5° Leo		
1461	Eckstorm, Hevelius, Sturm, Rockenbach, Cureus, Ming Shih	Gemini?		
1468	Pingré	Leo		
1471	Pingré	Libra		
1477	Eckstorm, Alsted, Rockenbach. Heber	?	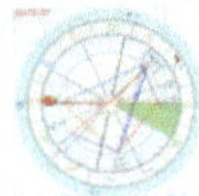	

There is just one technique of mundane astrology able to associate a celestial chart temporally well defined with a comet: the revolution of the year of discovery, namely the horoscope of the vernal equinox according to Arabic astrology or of the last luni-solar syzygy preceding the same equinox in the Ptolemaic tradition.[30] We have applied the technique in both versions to all the years of the comets, establishing a range of tolerance of half a house (about 1 h).

The result of this cross-correlation between comets and revolutions of the year is presented in Table 1, where two objects (the promising comets of 1468—the year of the meeting of Paul II and Frederick III—and 1477—the year of death of the Duke of Burgundy) are ruled out, leaving four possible dates for the star of the Sibyl.

[30] For details see G. Bezza, "Tolemeo e Abû Ma'shar: la dottrina delle congiunzioni Saturno-Giove presso i commentatori tolemaici", in *From Masha'allah to Kepler. The Theory and Practice of Astrology in the Middles Ages and the Renaissance*, London, The Warburg Institute, 13–15 November (2015), in press.

4.3 A Double Scenario

Among the four survived comets of the list, the Halley Comet has been selected only for its connection with Capistranus, but since the XVIth century attribution of the prophecy to a saint already dead at the time of the presumed vaticinium seems a complete invention,[31] the candidate may be considered of marginal interest and too early.

The 1460 object, on the other hand, has been chosen because its apparition coincides with the comet cited in the Capistranus prophecy, a fact that could be interpreted as a meaningful reminiscence of the original star. Nonetheless, reading the chronicles one finds only vague and inconsistent descriptions and no correspondences to Chinese observations, which is surprising for a comet that Boethius defines "*clarissimus*".[32] This comet seems to be a classical example of false observation, invented for the death of James II of Scotland and erroneously associated to the decease of Charles VII (actually happened an year later) (Rockenbach 1602). Its elimination leaves the most interesting candidates of the lot: the stars of 1461 and 1471, both consistent with the comets of the sibylline narrative, the first with the *Sybillen Weissagung*, the second with the *Proroctvì Sibyllino*.

This intriguing double scenario cannot be disentangled further, because each of the two comets has its advantages. The former, with Saturn in Capricorn during the syzygy preceding the vernal equinox,[33] was for example considered by Joachim Curaeus the cause of ravage in Bohemia[34] and could be identified with one of the two comets observed in China in that year in Gemini and Ophiucus.[35]

The latter was instead the extraordinary celestial body appeared in Libra during the winter of 1471–72 and observed among the others by Regiomontanus, Toscanelli and Bylica, which on the basis of this object and of the comet of 1468 predicted the death of George of Podebrady, and dangers for both the Pope and the Emperor.[36]

The comet of 1461 is very interesting because it could be associated to the war between two archbishops of the diocese of Mainz (the seat of the catholic primate in Germany, symbolized by the wheel in the image), protagonists of a bloody conflict begun in the same year that at the time should seem to reflect the Gamaleon Prophecy.

---

[31]The incongruence was already noticed by Stanislas de Liubenietski, *Historia Universalis Omnium Cometarum*, Leiden, (1681), p. 296.

[32]Boethius, *Historia Gentis Scotorum*, (1527), xvii.

[33]That is in its domicile sign, a configuration reinforcing the planetary strength.

[34]"*Fulsit cometa, qui proculdubio significavit cladem regni bohemiae, quae paulo post secuta est*", J. Cureus, *Gentis Silesiae Annales*, Wittenberg (1571), p. 166.

[35]See J. Williams, *Observations of comets from B.C. 611 to A.D.* (1640) *extracted from the Chinese annals*, London, 1871, pp. 315 and 354.

[36]A discussion of the two relevant *Judicia* is in D. Hayton, "Martin Bylica at the Court of Matthias Corvinus: Astrology and Politics in Renaissance Hungary", *Centaurus*, 49 (2001), pp. 185–198.

The comet of 1471–72, instead, has its strength in the prophecies about the future of princes that it generated, and especially in its connection with the last battle of Charles the Bold, whose fate was sealed also by the comet of 1468 according to many astrologers.[37]

The double scenario is in a sense very well represented metaphorically by the illustrated prophecy: depending on which figure one chooses, the dating rests on the lion under the foot of Frederick III, hence on Charles the Bold's fateful comets, or on the Mainz Wheel under the foot of Pius II, that is on the comet perhaps responsible of the "bellum moguntinum" of a decade before.

5 Conclusions

The Altino Prophecy that gave origin to the Pseudo-Capistranus vaticinium and inspired the "Oryetour Sydous" textual variation is a crucial and seminal episode of the narrative of the "star of the Sibyl", a tradition propagating across the centuries since classical times.

The evidences we have discussed so far seem to corroborate a bohemian or german origin of the image, while the astrological analysis dates its first textual commentary to 1461 or 1471, at least until new documents emerge from archives to clarify the issue.

Yet more than the seduction of puzzle solving, the case of this curious prophecy offers an extraordinary testimony of the inextricable connection between theology, astral divination, politics and natural science during the transition from Middle Ages to the Renaissance.

References

Ackermann Smoller, L.: Teste Albumasare cum Sibylla: astrology and the Sibyls in medieval Europe. Stud. Hist. Philos. Biol. Biomed. Sci. **41**, 76–89 (2010)

Alexander, P.J.: The medieval legend of the last Roman Emperor and Its messianic origin. J. Warburg Courtauld Inst. **41**, 1–15 (1978)

Barbieri, F.: Sibyllarum et Prophetarum de Christo Vaticinia, Roma (1481)

Bezza, G.: Tolemeo e Abû Ma'shar: la dottrina delle congiunzioni Saturno-Giove presso i commentatori tolemaici. In: Burnett, C., Greenbaum, D.G. (eds.) From Masha'allah to Kepler. The Theory and Practice of Astrology in the Middles Ages and the Renaissance. London

Boethius: Historia Gentis Scotorum (1527)

[37]For example in Alstedius: "*Duo cometae apparuerunt. ['…'] Carolus Burgundiae dux adversus finitimos eo impetus movit arma, ut multorum etiam annorum bellum excitaverit: quod finitum est ejus morte*" (cited in de Liubenietski, op. cit. 1681, p. 302). The two comets are probably conflated and represented in the Altino image by the lion and the scale, because they appeared respectively in Leo and in Libra.

Bonmann, O.: Zum Prophetismus des Johannes Kapistran, 1386-1456. Archiv Für Kulturgeschichte **44**, 193–198 (1962)

Cureus, J.: Gentis Silesiae Annales. Wittenberg (1571)

El Enigma de la Sibila. https://sites.google.com/site/omnedecus/Home/art/el-enigma-de-la-sibila

Geffcken, J.: Die Oracula Sibyllina. Heinrichs, Leipzig (1902)

Hakansson, H.: Tycho the apocalyptic: history, prophecy and the meaning of natural phenomena. In: Science in Contact at the Beginning of the Scientific Revolution, Prague, pp. 211–236 (2004)

Hayton, D.: Martin bylica at the court of Matthias Corvinus: astrology and politics in renaissance hungary. In: Centaurus, vol. 49, pp. 185–198 (2001)

Hélin, M.: Un texte inédit sur l'iconographie des sibylles. Revue belge de philologie et d'histoire **15**(2), 349–366 (1936)

Hind, A.M.: Early Italian Engravings. New York (1938)

Holdenried, A.: The Sibyl and her Scribes. Ashgate, Aldershot (2007)

Kristeller, P.O.: Der venezianische Kupferstich im XV. Jahrhundert. Mitteilungen der Gesellschaft für vervielfältigende Kunst **1907**, 1–16 (1907)

Lazar, M.: La legende de "l'arbre de Paradis" ou "bois de la croix". In: Zeitschrift für romanische Philologie, 76/1-2, pp. 51–52 (1960)

Levenson, J.A., Oberhuber, K. et al. (eds.): Early Italian Engravings from the National Gallery of Art. Washington (1973)

de Liubenietski, S.: Historia Universalis Omnium Cometarum. Leiden (1681)

McGinn, B.: Visions of the End: Apocalyptic Traditions In The Middle Ages. Columbia University Press, New York (1979)

McGinn, B.: Circoli gioachimiti veneziani. Cristianesimo nella Storia **7**, 19–39 (1986)

Neske, I.: Die spätmittelalterliche deutsche Sibyllenweissagung. Göppingen, Kümmerle (1985)

Parke, H.W.: Sibyls and Sibylline Prophecies in Classical Antiquity. Routledge, New York (1988)

Petrella, G.: La Pronosticatio di Johannes Lichtenberger: un testo profetico nell'Italia del Rinascimento. Udine, Forum (2010)

Pompeo Faracovi, O.: Gli Oroscopi di Cristo. Marsilio, Venezia (1999)

Reeves, M.: The Influence of Prophecy in the Later Middle Ages, pp. 343–346. Oxford (1969)

Rockenbach, A.: De Cometis. Wittenberg (1602)

Schreiber, W.L.: Handbuch der Holz- und Metallschnitte des XV. Jahrhunderts, 4, Leipzig (1926)

Secret, F.: Cornelius Gemma et la prophètie de la Sibylle Tiburtine, pp. 423–431. Revue d'histoire ecclésiastique, LXIV (1969)

Smahel, F.: Das zerbrochene Szepter des böhmischen Königs. In: Zur politischen Prasentation und Allegorie im 14. u 15. Jahrhundert, Munchen, Oldenbourg, pp. 39–75 (1994)

Thomas, A.: Anne's Bohemia: Czech Literature and Society, 1310–1420. University of Minnesota Press, Minneapolis (1998)

Vassilieva-Codognet, O.: The iconography of the last world emperor: a study of a late fifteenth-century prophetic image. Ikon **5**, 199–206 (2012)

Verkholantsev, J.: Ruthenica Bohemica. Ruthenian Translations from Czech in the Grand Duchy of Lithuania and Polonia. Wien, Lit Verlag (2008)

Williams, J.: Observations of comets from B.C. 611 to A.D. 1640 extracted from the Chinese annals. London (1871)

Wolf, J.: Lectionum Memorabilium et Reconditarum (1600)

Zurek, V.: L'usage comparé des motifs historiques dans la légitimation monarchique entre les royaumes de France et de Bohême à la fin du Moyen Âge. Ph.D. Dissertation (2014)

STARing the Sky in the Face: Recognizing the Constellations in a Sky Which Does not Have Any

Angelo Adamo

Abstract The analysis described here attempts to estimate the range within which we use our tendency to see a familiar shape in a disordered pattern (*Pareidolia*). The study starts with the proof that the stars visible to the naked eye are arranged following a Poisson distribution, a concept that I use to understand why, in the works of several artists, the stars appear so clustered that they form an excess in the number of possible constellations. This analysis should be considered only preliminary and will be completed soon by further investigation which I describe at the end of this paper.

1 Introduction

This paper has two "biological parents". One is a concept that has been gathering dust in my head since the period of my bachelor thesis: "In the night sky, the stars visible to the naked eye represent an excellent example of a Poissonian distribution of points".

The other parent is an analysis I could make, thanks to my fairly rich personal library (Adamo 2014), on how some of the most famous cartoonists have represented the night sky in their works.

In these artists' books, I noticed the tendency to draw the stellar points often so close to each other, as to return the idea of a sky which is much more populated by constellations than in real life. A big help to my analysis came from the fact that often in those boards, in addition to stars, the artist represented also the Moon, which has about a half degree angular width. When our satellite is present as in the Fig. 1, we can use it as an indicator of the width of a *pencil beam survey*: it tells us

A. Adamo (✉)
Osservatorio Astronomico di Bologna (INAF), Bologna, Italy
e-mail: angelo.adamo@oabo.inaf.it

A. Adamo
Università dell'Insubria, Como, Italy

A. Orlando (ed.), *The Light, The Stones and The Sacred*,
Astrophysics and Space Science Proceedings 48,
DOI 10.1007/978-3-319-54487-8_14

Fig. 1 A frame drawed by G. Alessandrini

what the artist's perception of an angular extent of about half a degree is, and thus allows us to measure the relative distances of stellar points drawn in the same board.

Understanding why this happens seemed very challenging to me, and I have found that there are several reasons for the presence of such a trend in many different authors' boards, some easier to understand than others.

An analysis of some of them, for example, has shown that the cartoonist, after drawing a white dot (the star) on a black background (the sky), simply moved his hand fast and slightly. This is indicated by the fact that the stars drawn by those cartoonists often are all of the same magnitude, which tells us how they used the same pen, or a *Photoshop* brush of the same thickness.

I claim this from personal experience and also because other professionals confirmed my idea: usually this happens in the attempt to shorten the delivery time of a work whose primary (or even secondary) purpose certainly was not to educate readers on the true distribution of stars in the sky.

The editorial times are always so tight that the artist is always bound by the mandatory terms under its contract, often at the expense of the quality of some boards, if not of the entire publication.

This does not apply to the famous cartoonist who can afford frequent violations of the terms of his contract, and therefore can spend more time on the effort to represent that piece of the world with a certain realism, with exactly the same zeal with which he has always studied and represented a horse, a palace, the folds of a dress, an airplane, etc. My question "why are you not as keen to correctly depict the stars, which often are the natural background of the scenes you draw?", I think, is more than legitimate, and it would be interesting to investigate what are the artist's mental processes when selecting the elements of a scene—other than the obvious ones such as, for example, the main characters of the story—on which to focus the pictorial attention.

Another possible explanation for this trend to depict an excess of constellations, shown by cartoonists and by some painters, may derive from the real existence of a stronger correlation between the dots in the sky which we scientist have not noticed.

So it may be that, in the absence of a real analysis of the distribution of stars visible to the naked eye, we tend to believe erroneously that their position on the celestial sphere is random, while the eye of the artists, who I usually trust, spontaneously captures a real correlation between those light points.

It is unlikely that astronomers, who are used to seeing always and only the same constellations, may express the need to see others; but a cartoonist and, in general, an artist who is not also an amateur astronomer, could perhaps experience this kind of impulse and follow it. If so, I would like to understand where this impulse comes from and if, by chance, it can be explained just by the fact that the stars turn out not to be distributed in a random way.

2 Looking for the Real Distribution of Visible Stars

In a search of the literature to see whether an analysis of the distribution of stars in the sky has ever been carried out, I have not found anything significant other than phrases like "assuming a random distribution for the stars" or quick, cryptic references to the fact that things are just as the authors had stated earlier in the paper: according to all the authors I have consulted, stars in the sky occupy positions which follow a Poisson distribution. I then decided to check out this assumption using a program written by Alberto Cappi[1] to calculate the *two points angular correlation function* $w(\theta)$ for groups of galaxies. The choice to use only the *two points angular correlation function* $w(\theta)$ and not the most popular $\xi(r)$, the *spatial two points correlation function* is due to the fact that, observing the sky with the naked eye and seeing the constellations that enter our field of view, we always ignore the distances of those stars from our position.

One should be reminded that the *two points angular correlation function* $w(\theta)$ is a statistical tool which, starting from a point chosen in the ensemble, allows us to calculate the excess of probability of finding coplanar neighbors of the given point, at a previously established angular distance θ. If this function returns an average value of 1, it means that the points are not randomly distributed, presenting quite a strong tendency to cluster. If instead the $w(\theta)$ is, on average, equal or close to zero, it follows that the set of points is distributed at random. Finally, if the correlation function has a negative value, we will say that the whole set is not correlated.

The goal of this analysis is to measure how the constellations are formed by stars with a high probability of being seen as connected because, projected on the sky, they are really close to each other; they could be so close to stimulate in us the belief that they really draw the forms to which we have been referring for about 5000 years when speaking of "astrological signs". Each time it is run, the program applies the procedure starting from each of the N points of the ensemble considered

[1]Osservatorio Astronomico di Bologna (INAF); Laboratoire Lagrange (Observatoire de la Côte d'Azur, Nice).

and, once the *i-th* analysis is conducted for a certain value θ_i of the angular distance and has ended, starting the $i + 1$ cycle, the program increases the former angular distance θ_i adding to it a fixed quantity chosen by the user. This way the procedure goes on until—and here the analysis ends—it reaches the angular size of the total area occupied by the whole set of points. At each step, the program then applies a comparison with a fictitious catalog of ideal star points distributed in a very random way—this means that sure it has a $w(\theta)$ equal to zero, unambiguous proof of it being a Poisson distribution—on a surface area equivalent to that of the sky. The final calculation of the correlation of the points (stars) belonging to the real catalogue is then performed using the following relationship:

$$w(\theta) = [N_{SS}(\theta)/N_{RR}] - 1$$

where N_{SS} is the number of star-star pairs separated by a given θ angle, and N_{RR} is the number of pairs of fictitious star points distributed randomly and separated by the same angle.

3 The Catalogue

I have then asked Alberto Cappi to use his program on a star catalog, the BSC5P, which is the fifth edition of the *Bright Star Catalogue* (Hoffleit and Warren 1991), restricting the analysis to the stars with apparent magnitude less than or equal to 6 (strictly speaking, all and only the stars visible to the naked eye). For the purposes of this first work, I did not need to consider the apparent magnitude differences between the stars of the catalog, focussing my attention only on their position. It is my intention to soon expand this work using this same program which our colleague Federico Marulli has modified, allowing us to weight the correlation with the values of the various known apparent magnitudes and checking whether or not there is a higher correlation between stars of different apparent magnitude.

To perform the analysis, I have considered only the galactic coordinates of any single star, dividing them in the following four different sets:

(1) stars belonging to the northern galactic hemisphere ($0° \leq BII \leq 90°$, total number of objects = 2352, density (N/area) = 0.19600E + 03);
(2) stars south of the galactic plane ($-90° \leq BII \leq 0°$, total number of objects = 2752, density (N/area) = 0.22933E + 03);
(3) stars positioned to the north of the strip of the galaxy ($30° \leq BII \leq 90°$, total number of objects: 942, density (N/area) = 0.78500E + 02) (I have assumed an approximative thickness of the disc of Milky Way as seen from our position equal to 60°, 30° for each hemisphere);
(4) stars south of the galaxy strip ($-90° \leq BII \leq -30°$, total number of objects = 986, density (N/area) = 0.82167E + 02).

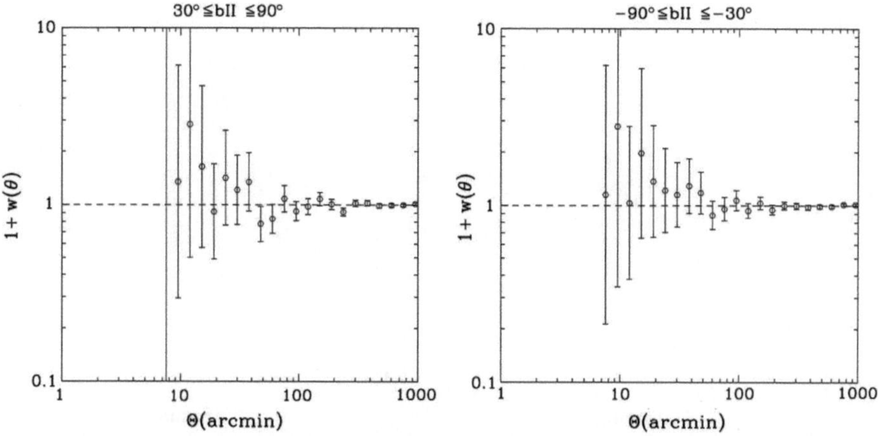

Fig. 2 *Left* correlation of stars positioned to the north of the strip of the galaxy (30° BII ≤ 90°); *Right* correlation of stars south of the galaxy strip (−90° ≤ BII ≤ −30°)

I have done this with the purpose of quantifying the gravitational noise (in this study, I like to regard it this way, but we know that it is a "signal") given by the presence of the galactic plane, especially in the milky band that we see in the sky. It is well known that the Milky Way strip is part of the arm of the galactic disk of stars to which our Sun belongs, observed from a distance that enables us to appreciate its thickness in perspective: an area of the sky crushed in a few degrees of galactic latitude. The very fact that many stars belong to that strip, which is a few tens of degrees thick, causes the eye to see, if not always constellations, at least a correlation among those stars which are really close to each other.

The result of the analysis carried out across the sky, excluding the stars contained in the strip of the Milky Way, can be assessed from the graph shown in Fig. 2. The correlation between the stars is appreciable only at angular distances of the order of a few tens of arc minutes. These are distances similar to the size of the smaller asterisms, many of which do not contain stars of particular brilliance and are also among the most difficult to note: the *Equuleus*, the *Pleiades*, the *Arrow*, the *Dolphin*. As can be seen from the graph, observing how the w(θ) function goes quickly to zero, when I write "appreciable", I simply mean that, even if only slightly, the correlation function assumes non-zero values. As you can see, as the range increases, the sampled correlation decreases rapidly, and already on scales of the order of a degree it is practically zero. Another consideration that emerges from the study of Fig. 2: there is no substantial difference between the two hemispheres above and below the center line that, running through the Milky Way, defines the galactic plane from which I considered stellar positions.

If you decide to include in the correlation analysis also the stars belonging to the galactic plane strip, you get, as expected, a slightly stronger correlation (Fig. 3). In any case, the results do not differ much from the previous case in which the Milky Way had been excluded.

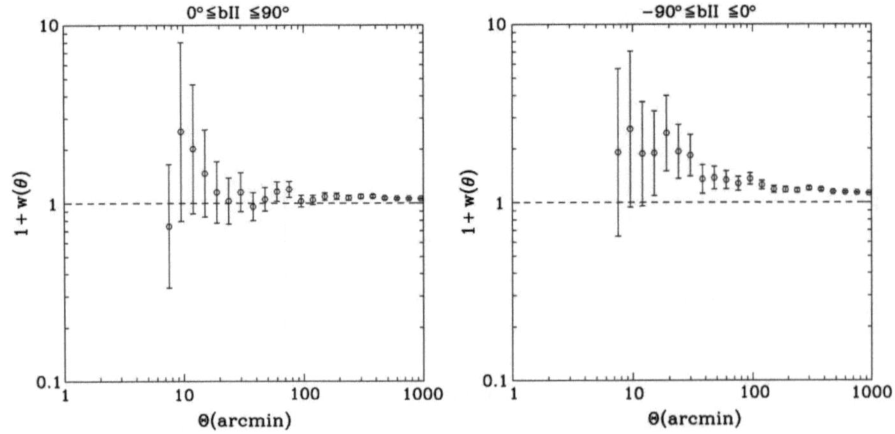

Fig. 3 *Left* correlation $1 + w(\theta)$ of stars belonging to the northern galactic hemisphere ($0° \leq$ BII $\leq 90°$); *Right* correlation $1 + w(\theta)$ of stars south of the galactic plane ($-90° \leq$ BII $\leq 0°$)

By overlaying the results of both correlation studies (Fig. 4), it is easily seen from the graph that at the great angular scales of the order of tenths of degrees that characterize the size of the largest and often more famous constellations, the stars do not appear correlated. This result, obtained starting from the idea of assessing the plausibility of the heavens drawed in some works of cartoonists and painters, tells us that these artists are "victims", like anybody else, of a perceptual error widespread in our species, that leads us to overestimate the correlations between stars in the sky.[2] I now assume that a possible assessment of our ability to see the correlation between points can be approximatively measured using an angle slightly larger than the resolving power of the eye (≈1′).

In the past, to understand if you had a sight of 20/20, you would have to try to resolve by naked eye the two stars Mizar and Alcor which, as we now know very well, are 11′48″ apart. As Bohigian in his paper demonstrates (Bohigian 2008), that test is entirely equivalent, if not even more reliable, to the classic Snellen test we face every time we go to an eye clinic. As proof of the fact that, as a species, humans cannot avoid attaching special significance to manifestations of reality that appear significant, we note that various cultures have regarded those two very close

[2]Among those drawn by cartoonists, painters and illustrators, the best skies I found are those for which the artist used a technique that renders a true Poisson distribution of the painted stars. This is an old trick that consists in using a toothbrush previously dipped in a layer of white paint. By suspending it over the dark background meant to represent the night sky, the bristles are rubbed with a thumb, causing a splash of paint drops that will distribute randomly on the paper or canvas. This technique has the added benefit of creating also certain randomness in the distribution of magnitudes of the various points. It is used more by illustrators than by cartoonists, because it requires large areas (the traditional comic strip occupies small spaces), and/or appropriate masking of the other elements of the image on which no drop must fall.

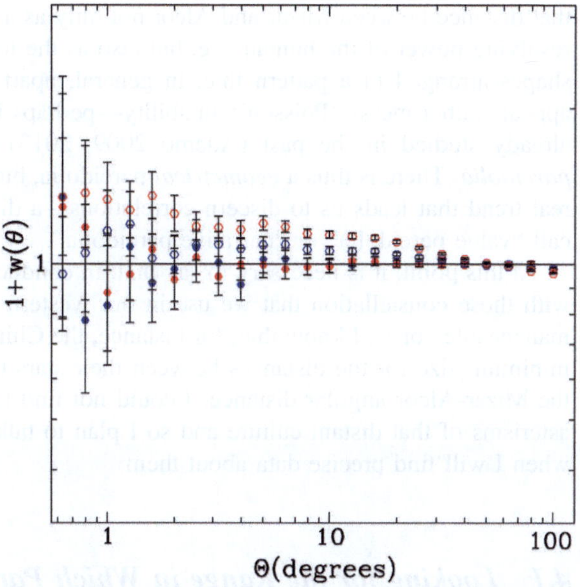

Fig. 4 General correlation $1 + w(\theta)$ between the angular positions of visible stars. *Blue circles* $0° \leq BII \leq 90°$; *blue dots* $30° \leq BII \leq 90°$; *red circles* $-90° \leq BII \leq 0°$; *red dots* $-90° \leq BII \leq -30°$

stars as (a) a squaw with her baby on her shoulders; (b) an indian hunter carrying a saucepan; (c) a small rider on his horse's rump.

Having noted this strong tendency to see histories and meanings even in a minimal segment like the one seen joining the two stars of the so called *Big Dipper*, I was drawn to consider them as important asterisms, as well as the most extensive and well-known constellations. Moreover, the aforementioned *Dolphin* or the *Pleiades*, even if very small stellar clusters, confirm this *attaching tale*-tendency of the human brain. Moreover, if this still is not convincing enough, let us remember that, for example, the constellation of *Aries*, larger than those just mentioned, is built around a simple junction between two very simple segments. This graphic simplicity has never been a problem for those who insisted (and continue to do so) on fleshing out that minimum skeleton, imagining around it the development of a curious cross-breed between a goat and a fish, making it a full sign with astrological meanings ranging well beyond the one expressed graphically.

4 Pareidolia

Considering all the graphs showed before, we also note that at the scale of about one-sixth/one third of a degree they all shows that our eye actually could catch a real correlation doomed to flounder at higher scales. Remembering the meanings given to those two stars and the medical use that ancient people have done of the capability of a human eye to resolve those two stars, I would propose to consider that distance between Mizar and Alcor not only as a fairly accurate measure of the resolving power of the human eye, but also as the lower limit of our ability to see shapes arranged in a pattern that, in general, apart from some small deviations, appears rather messy (Poisson): an ability—perhaps I should say a "trend"—I have already studied in the past (Adamo 2009, 2013) which goes by the name of *pareidolia*. There is thus a *geometrical* pareidolia, but also—and perhaps this is the real trend that leads us to discern correlations—a different kind of it that I would call "value pareidolia" or "narrative pareidolia".

At this point, it is necessary to say that, in conducting this analysis, I dealt only with those constellation that we use in our western culture to divide the sky into manageable zones. I know that, for instance, the Chinese constellations can reach a minimum size for the distances between their stars that, they say, are shorter than the Mizar-Alcor angular distance. I could not find reliable data on the size of the asterisms of that distant culture and so I plan to talk about them in a future work when I will find precise data about them.

4.1 Looking for the Range in Which Pareidolia Acts

If you accept my idea of considering at a first approximation the angular distance between Mizar and Alcor as a measure of the minimum distance at which the pareidolia acts, the next question comes in a natural way: what is the maximum size at which our visual system perceives a correlation between related and/or even uncorrelated objects? Studying the literature, I found no answer to this simple question. There are careful studies of how pareidolia acts in the complicated case in which the brain tries to recognize a face to which a pattern of random Gaussian bivariate blobs has been superposed (Andrews et al. 2002; Chauvin et al. 2005; Gosselin and Schyns 2003; Howard and Rogers 1996; Liu et al. 2010, 2014; Rieth et al. 2011). Another trend in these studies concerns how we grasp the differences between the different characteristics of human faces, but I have not found anything quantifying the maximum angular dimension within which our brain creates fictitious correlations between objects that are not related.

At this point, given that (1) humankind never showed a tendency to consider a *meta-constellation* of angular size comparable to that of the portion of the sky visible in a single glance and, by contrast, (2) humans have always shown a marked tendency to see in a large sky field many of constellations of different angular sizes, ideally fragmenting the field of view, in order to understand which is the range of action of our brain while it "sees" in the sky the objects that from Figs. 1 and 2 we know are not real, I will just consider the larger constellations that have a size smaller than or equal to our visual field.

5 Field of View of the Human Eye

For the non-trivial evaluation of what the extensive human field of vision is, I was advised to read the following passage extracted from the book "Binocular vision and stereopsis":

> The monocular visual field of the stationary eye extends about 95° in the temporal direction and about 56° in the nasal direction (Fischer and Wagenaar 1954). The total visual field is the solid angle subtended at a point midway between the two eyes by All Those points in space visible to either eye or both. It extends laterally about 190° in humans When the eyes are stationary and about 290° if they are allowed to move. If the head moves on the stationary body, the total visual field extends through almost 360°. The binocular visual field is the portion of the total field Within Which an object must lie to be visible to Both eyes for a given position of the eyes. The binocular visual field is flanked by two monocular sectors Within Which objects are visible to only one eye. Each monocular sector extends about 37° laterally from the temporal boundary of the orbital ridge to the boundary of the binocular field at infinity. Each monocular visual field is the sum of the binocular field and the monocular sector For That eye. The left and right boundaries of the binocular field, formed by the nose, are about 114° apart When the eyes converge symmetrically and less When They converge on an eccentric point. The horizontal extent of the binocular visual field in the 3-month-old human infant Has Been estimated as 60° and that in the 4-month-old infant as 80° (Finlay et al. 1982). With the eyes in a straight ahead position, the upper boundary of the binocular field, formed by the orbital ridges, extends about 50° above the line of sight. The lower boundary extends about 75° below the line of sight. The blind spot, the region where the optic nerve leaves the eye, is devoid of receptors. The projection of the blind spot in the visual field is about 3° in diameter and about 12°–15° falls into the temporal hemifield. Hence, there are two islands Within the monocular binocular field, one on each side of the point of convergence. The field of binocular fixation is the area Within Which binocular fixation is possible by moving the eyes but not the head (Sheni and Remole 1986).

From reading this passage and from a comparison of the numbers mentioned in it, the average amplitude, minimum and maximum of forty-eight ancient constellations, we see clearly how they are perfectly contained in the solid angle subtended by our gaze. To obtain the amplitudes of the constellations, I have used the atlas by Eckhard and Uwe (2006) that for every constellation, gives the coordinates of the "higher star and of the lower star", as well as of the "rightmost" and "of the leftmost". By calculating the differences between δ_{Max} and δ_{Min} and between AR_{max} and AR_{min}, multiplying the latter difference by the cosine of δ_{Min}, I found the "real"[3] amplitudes of fifty ancient constellations (originally there were forty-eight, but the *Argo ship,* perhaps considered too big to be seen at a glance, was divided by Nicolas Louis de Lacaille (1713–1762) into three different parts: *Puppis*, *Carina* and *Vela*). The maximum amplitude that occurs in RA is $\Delta AR_{max} = 116.86°$, the minimum is 6.2. The maximum amplitude is $\Delta\delta_{max} = 56.5°$ while the minimum is 5.5. The average sizes instead are $\langle AR \rangle = 36.38$ and

[3] I could not use the areas assigned to each constellation by Dalporte because they exceed the dimensions of the figures drawn in the sky using the stars. For obvious reasons, I am interested only in the areas included in the starry perimeters.

$\langle\delta\rangle = 28.28°$. There are some exceptions. For instance the *Hydra* constellation, by far the largest one, is not completely visible at a glance because, if one starts observing it at our latitudes, it turns out that it stretches below our horizon, where we just cannot see anything from here. In this case, ancient sailors and merchants have implemented an ideal extension of this constellation that, seen from here, appears to end at the horizon. Travelling south, they realized they could consider *Hydra* larger than previously known, due to the presence of other southern stars that could be easily considered as its extension. In general, a generic constellation is perfectly contained in the visual field of the observer, who often sees more than one simultaneously.

6 Toward the High Limit of Pareidolia

At this point, I would try to propose an upper limit to the physical range within which our pareidolia acts: I propose it is equal to the size of a theoretical constellation that is as wide as the widest of the known ones, 116.86°, and as high as the one which has the largest declination range, 56.5°. If one wishes to consider that not everybody has the same visual skills, one may assume that in reality the upper limit of our ability to see correlations even where there are none, is more or less given by a constellation that has an average size, 36.38° wide in right ascension and 28.28° in declination.

7 Future Developments of the Research: Percolation

It is well known that when we look at a face, at an object, at the sky, etc., even though we are sure to keep our eyes well fixed on what we observe, we make a series of tiny, fast movements called "saccades" which put the different parts of what we observe in correspondence with the *fovea*, the most sensitive part of our inner eye. In other words, what we do is to unite the points of an object on which we focus our attention to reconstruct its general appearance. To understand how our brain works when we are sure to gaze at the constellations, I want to continue the present analysis again thanks to another of Alberto Cappi's programs, that analyzes a set of elements implementing the so-called "percolation": it starts from any point selected in the ensemble and then it moves, by a step of amplitude previously decided by the user, in the direction of the first star that falls within the circle of radius equal to the step itself. Our brain, once the eyes move (saccades) between various stellar points, perform a continuous comparison between the geometry of the brightest star distribution and various mental images that it possesses in its memory. In the case of the ancient constellations these images are bulls, bears, lions, and all the other animals, characters and objects we usually see in the sky. By analogy, using the percolation, the program that I want to use will have to perform a comparison between the points of a catalogue and a collection of simple templates such as rectangles, circles and triangles.

Percolation will need to be weighed in the sense that, if two stars of different brightness fall in the circle of radius equal to the step value, the program will choose to go to the brighter one. In this way, what I expect is to find a value of the step (sensitivity) that allows the program to find shapes between the points of a star catalog of sizes comparable to those of ancient constellations listed by Ptolemy in his Almagest. Such a result could have interesting implications in the study of the so-called *pattern recognition problem*, which refers to the problem of figuring out how to make a computer system able to recognize a particular shape. A natural application of this research could then be its implementation in optical systems of industrial machinery.

8 Conclusions

In this paper which more or less travels on the same path traced by Bohigian's, I have used the sky not as an object of study, but as an instrument to understand our "human universe". Perhaps everytime we discover that astronomy could help us in understanding our society, our history, our physiology etc., we can talk of a sort of "medical astronomy" or about a "self-astronomy" or even a "social astronomy".

I have analyzed the possible physical range within which the visual pareidolia occurs, starting from the act of viewing which uses the "eye" medium, a powerful instrument but one which, as we all know, is also affected by some limitations. At this point, I would like to investigate if there is also a "tactile pareidolia" in blind observers, and if so in which range it acts: I suspect that visually impaired people divide the field in which they move in touchable, recognizable and comparable fractions (I would call them "object constellations"). In these space fractions they look for recognizable (archetypal?) patterns of objects the same way we manage the visual field in handy fractions of the sky (constellations) looking for well known shapes.

Continuing by analogy, I believe that we are able to use also a sort of *neural pareidolia* which allows us to connect together apparently unrelated thoughts: sometimes it happens that someone catches a possible correlation between different, distant concepts, while others continue to consider them well separate. This kind of possible *neural pareidolia* clearly merges with the more popular *concept* of *intuition* or in that of *intelligence,* and I strongly trust in the progress of neuroscience that, I hope, will soon tell us which, if any and if measurable, is the upper limit of what we can imagine, perceive, understand of the world in which we all come together. I think that our maximum capability to connect far concepts will be found to be comparable in some way to the measure of our skull dimensions which impose an upper geometrical limit to the maximum distances between brain neurons that, after receiving an external stimulus, activate together or are directly connected to each other by a cause-effect relation. Another possible measure could be in the number density of neurons working together on the same stimuli, and a first "proof" of the existence of such a limited capability could be present in our way of designing the chips of our computers, making them more and more dense of "wires", as well as in our tendency to pull the web to extreme distances and local ramifications. This happens maybe because humans tend to group together in the attempt to extend the limited physical size of their individual brain (and thoughts), creating connections between their ideas and opinions that otherwise would be limited by lives spent staying alone. We all share ideas and values and everyone is at the same time a generator, a container, a vehicle and an antenna of concepts. History teaches us that interacting people don't appear always as guided by some form of superior intelligence. Perhaps, like the perception of constellations, there is a maximum size of human ensembles within which one can reasonably hope to see a real growth of the democratic values and of the behaviour of each component. In contrast, exceeding that limit, we always measure a worsening: perhaps the "field of view" of the entire human assembly is too large to work in a good way and we all need to break down the society into states, regions, collectives, families, parties, clubs, etc. If true, this could explain why some are still racist while others simply believe in some form of "federalism".

Perhaps machine evolution one day will frighten us so much to stimulate us to learn how to live in a real democratic way with all other humans on the planet. Another possibility is that we will expand the size of our "democratic field of view" learning to think on a larger social scale only when we will meet an alien civilization: this extraordinary event could stimulate a social dynamic—something similar to a fractal—convincing us to use our Darwinian behaviour against the "others" and not against ourselves. In that case, we will all have a single, well known face but pareidolia will not help us to recognize another specie's face.

Acknowledgements I thank Professor Marco Costa of Bologna University for providing me with the text regarding the width of the human field of view, and Dr. Carlo Alberto Zotti for sending me information on night vision and for introducing me to the paper by Bohigian.

Thanks also to two of my colleagues: Federico Marulli, who has offered to help in the continuation of this research, and Giovanna Stirpe, who has kindly translated this paper in a correct way (it was originally written in Italian). Finally I have to thank Alberto Cappi for sharing with me his time, his advice, his knowledge and above all his program, in the name of a friendship of which I am proud, and that has now lasted for more than 15 years.

References

Adamo, A.: Pianeti tra le note. Springer (2009)

Adamo, A.: …e tornammo a riascoltar le stelle, Giornale di Astronomia, n. 3 (2013)

Adamo A.: Stelle e strisce. I fumettisti incontrano l'Astronomia. Bachelor Thesis, Academy of Fine Arts, Bologna (2014)

Andrews, T.J., Schluppeck, D., Homfray, D., Matthews, P., Blakemore, C.: Activity in the fusiform gyrus predicts conscious perception of Rubin's vase–face illusion. Neuroimage **17**, 890–901 (2002)

Bohigian, G.M.: An ancient eye test-using the stars. Surv. Ophthalmol. **53**(5), 536–539 (2008)

Chauvin, A., Worsley, K.J., Schyns, P.G., Arguin, M., Gosselin, F.: Accurate statistical tests for smooth classification images. J. Vis. **5**(9), 659–667 (2005)

Eckhard, S., Uwe, R.: L' atlante delle costellazioni, Zanichelli (2006)

Finlay, D. et al.: Detection of moving stimuli in the binocular and nasal visual fields by infants three and four months old. Perception, **11**, 685–690 (1982)

Fischer, F.P., Wagenaar, J.W.: Binocular vision and fusion movements. Doc. Ophthalmol. **7**(8), 359–391 (1954)

Gosselin, F., Schyns, P.G.: Superstitious perceptions reveal properties of internal representations. Psychol. Sci. **14**(5), 505–509 (2003)

Hoffleit, D., Warren, Jr., W.H.: The bright star catalog, 5th edn (Preliminary Version) (1991)

Howard, I.P., Rogers, B.J.: Binocular Vision and Stereopsis. Oxford University Press, Oxford (1996)

Liu, J., Li, J., Feng, L., Li, L., Tian, J., Lee, K.: Seeing Jesus in toast: neural and behavioral correlates of face pareidolia. Cortex **53**, 60–77 (2014)

Liu, J., Li, J., Zhang, H., Rieth, C.A., Huber, D.E., Li, W., et al.: Neural correlates of top-down letter processing. Neuropsychologia **48**(2), 636–641 (2010)

Rieth, C.A., Lee, K., Liu, J., Tian, J., Huber, D.E.: Faces in the mist: illusory face and letter detection. i-Perception **2**(5), 458–476 (2011)

Sheni, D.D., Remole, A.: Field of vergence limits. Am. J. Optom. Physiol. Opt. **63**, 252–258 (1986)

The App STAR: An Important Instrument for Creating the First Spreading Thematic Museum on the Archaeoastronomy in Rome

Chirri Maurizio, Lombardi Livia, Ruggiero Ludovica and Fabrizi Lucilla

Abstract STAR is an important instrument developed for creating the first spreading thematic museum dedicated to the archaeoastronomy in Rome. This is the first italian App for smartphone and tablet able to guide the tourist on a journey to discover mithraea, sundials and Astronomy related places of the Capital. Although they attract millions of tourists, the astronomical significance of these sites is not valorized in most cases. For these reasons, STAR offers the opportunity to discover an original and unconventional side of Rome through an innovative technological support. This App aims to invite the foreign tourist as well as the Roman citizen in rediscovering the city, the neophyte as well as the expert of this topic. With STAR it is possible to personalize the itinerary, and to choose the number and type of sites to visit. At present, the App STAR includes ten specific astronomical sites. To communicate contents with pathos, the App reveals the original appearance of the sites and brings to light hidden details by using the virtual augmented reality. Virtual reconstructions and augmented reality were developed for two sites, the Horologium Augusti in Campo Marzio and the Basilica di S. Maria degli Angeli. Overall, this contribution gives added value for museological and museographical areas.

1 Introduction

S.T.A.R., an acronym meaning "Sites for Astronomical Tourism in Rome", is the first Italian application for smartphone and tablet entirely dedicated to archaeoastronomical tourism. Many sites in Rome own a double cultural value: artistic and astronomical. Unfortunately, the astronomical value of these sites is not valorize,

C. Maurizio · R. Ludovica (✉) · F. Lucilla
Department of Science, University of "Roma Tre"—Civico Museo Geopaleontologico "Ardito Desio", Rocca di Cave, Rome, Italy
e-mail: ludovica.ruggiero@gmail.com

L. Livia
Department of Earth Science, University of Rome "La Sapienza", Rome, Italy

A. Orlando (ed.), *The Light, The Stones and The Sacred*,
Astrophysics and Space Science Proceedings 48,
DOI 10.1007/978-3-319-54487-8_15

but only their historical and artistic value are known by the millions of tourists. The aim of STAR is therefore to offer the opportunity to discover an unexpected and original aspect of Rome through an innovative technological support, simple and interactive, which STAR wants to be.

The research to gather the data for the development of App were conducted in the period between 2004–2015 during the course of Museology (Master Degree in Technology for the Conservation and Restoration of Cultural Heritage, University of Rome “La Sapienza”). This course included a short course of museography and museology applied to archaeological sites, monuments and artworks connected to the History of Astronomy and the Cultural Astronomy in Rome. The students produced short thesis and museographic exercises on several sites connected to the history of Astronomy and Cultural Astronomy in Rome; such as historical sundials, obelisks, mithrea, monuments with astronomical meanings, museal collections with involved specimens and exhibits.

The occasion to realize this project has arrived whit the selection of the Regione Lazio, especially targeted for young creative people and aimed to the developing of innovative App. Each winning young participant had the occasion to be connected and work with a company for the development of the App. The project S.T.A.R. (Sites of astronomical tourism in Rome), presented by our former student Dr. Livia Lombardi, in collaboration with Dr. Ludovica Ruggiero and Dr. Lucilla Fabrizi, was chosen by Mobile Soft s.r.l. and classified first on the list of this selection. In this work, we report the results of our research, some content available in the app and guide lines for using our software.

2 Development of the Project: Didactic and Spreading Purposes

Particular museal tipology is represented by the Spreading Thematic Museum. In this kind of museums exists a theme, which is developed in several different locations, maybe small or minor, not necessarily close to each other, and they don’t attract many visitors. The app S.T.A.R wants to be the necessary instrument to connect every sites and to make the spreading museum accessible to the tourist. The aim of this project was to realize the basis for a spreading thematic museum about archeoastronomy in Rome using our research and the collected student’s work. The idea has started from the holders of the course of Museology, Prof. Luigi Campanella (until 2014) and later Prof. Giorgio Manzi. The realization of an app entirely dedicated on the archaeoastronomy in Rome offers the opportunity to realize the widespread museum on the subject. At present, the app STAR includes ten specific sites with astronomical and cultural meaning (Fig. 1).

We studied the ancient cartography of our city, especially the maps of Etienne Duperac (1535–1604) (Fig. 2), and the great map of G.B. Nolli (1692–1756) (Fig. 3). In this way, we have utilized a particular shape of Rome, formerly

Fig. 1 Students from the course of Museology, measuring the meridian line of the sundial "Horologium Augusti", found in via Campo Marzio 48 (June 2015)

identified by the studies and researches of the famous archeologist R. Lanciani (1845–1926) and, later, by the architect and urbanist P.M. Lugli (1923–2008). Their works hypothesized that, in the imperial age, some civil architectonic complexes would have built in specific sites to promote some morphological features of our city and to draw an "altera forma urbis", like a star (Fig. 4). In particular, P.M. Lugli thought that a long thread connects (links), some hard meaning's lines, of the Historia Naturalis of Pliny the Elder (L. III, parr. 65–67 cit. op.), to the great and magnificent architectonic arrangement of Rome, by Domenico Fontana (1543–1607) for the Jubelee of the 1600 (Lugli 2006). Such hypothesis has been perfect for our aims, giving to the thematic museum on Astronomy in Rome an enhanced cultural suggestion. For all of the sites tourist and cultural information in two languages (English and Italian) are available. The attention to internationalization is

Fig. 2 Etienne Duperac's map relative to the area of Campo Marzio, including a reconstruction of the ancient sundial (from "Urbis Nova descriptio")

essential because this tool also wants to involve the foreign tourists who, in many cases, have a marked sensitivity towards the astronomical culture. In this way, everyone can create their own journey in discovery of roman archaeoastronomy

Fig. 3 Representation of Claudius Ptolemy's Ekumen. G.B. Nolli's large map, visually suggesting the "alter shape" of the city

exactly as they are within a single museum. For two sites, the Horologium Augusti in Campo Marzio and the Basilica di S. Maria degli Angeli virtual reconstructions and enhancing reality, based on research and museological- museographics studies, have been developed.

3 Work's Methodology: The Case of the Horologium Augusti and the Meridiana Clementina

To create the virtual reality for the reconstruction of the major solar sundial in the classic ancient world, the Horologium Augusti builded in 10 BC in Campo Marzio, with the Psammetico's obelisk, we have chosen between two reconstruction models, proposed by two different scholars (Plinio 1982). E. Buchner was an archeologist of The Deutsch Archeologische Institute; he dipped the area in Via Campo Marzio, at the end of the 70s, and found a piece of the ancient bronze meridian line of the sundial. He proposed that the ancient device would have had a large travertine platform. This platform had a certain number of hour lines (variable

Fig. 4 Reconstruction of the Star shape of Rome in P.M. Lugli's map

between 7 and 11), the monthly hyperboles and the meridian bronze line with metallic daily lists (Fig. 5).

The other reconstruction model was proposed later by Heslin (2007), and it is composed only by the obelisk, his pedestal, and the meridian line with narrow travertine slabs. This model is today positively considered by an interdisciplinary team of researchers of Indiana's University, leads by Prof. Bernie Frischer. A solution of the problem would come from a new set of geological prospections, by non-invasive methods of georadar, searching the remains of the ancient platform, probably situated at 6.5–8.5 m under the street level. Actually all the local area has been largely and deeply altered by several works, and the remains of ancient large travertine platform could be entirely removed. Some student's works have recovered historical and

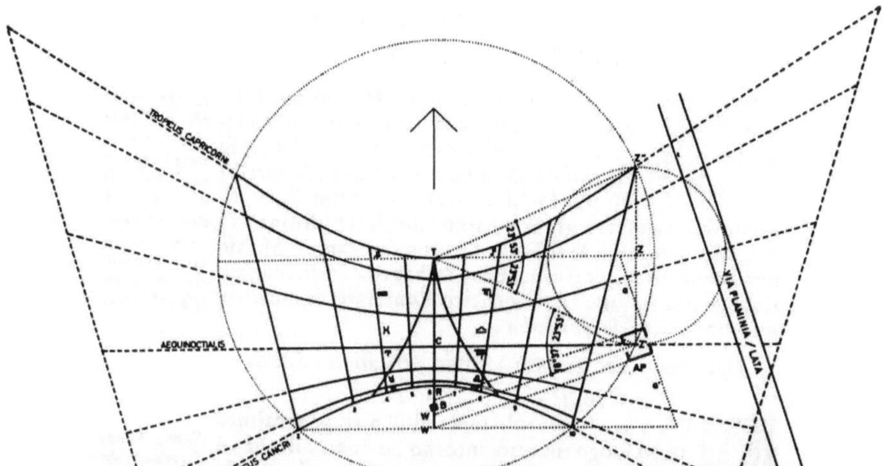

Fig. 5 Representation of the sundial "Horologium Augusti" with its wide hourly platform, in the reconstruction of E. Buchner

literary information about the first three discoveries of the Sundial (the Obelisk fallen at the ground, probably between the IX–XI century), occurred about in 1480, 1550, and 1748. A wide collection of citations and data, was recovered by the papers of the so called "antiquarians" in Rome between the XV–XVI century: humanists, archeologists, draughtmans and engravers, such as Flavio Biondo (1392–1463), Pomponio Leto (1428–1498), Andrea Fulvio (1470–1527) (Fulvio 1588), Bartolomeo Malviani (1488–1566) (Malviani 1632), Lucio Fauno (1501–1560), Pirro Ligorio (1513–1583), Michele Mercati (1541–1593), Alessandro Donati (1584–1640) (Donati 1648), A.M. Bandini (1725–1803) (Bandini 1751). Some authors had a direct knowledge of the digs, with specific witnesses of unknown details until that moment, indeed some lines were the only available classical source, collected in the Historia Naturalis of Pliny the Elder, (L. XXXVI, parr. 71–73 cit. op.) (Fig. 6).

Two authors, particularly, Pomponio Leto in the "Excerpta a Pomponio dum inter ambulandum cuidam domino ultramontano reliquias ac ruinas urbis ostenderet", and Andrea Fulvio in the "Dele antichità di Roma", give us a straightforward report of the digs, occurred about in 1480, for works in some district's houses (Pomponio 1953). These witnesses describe a large travertine platform with a certain number of hour bronze lines, with inscriptions and musive works of the winds. These and other informations were used by A. Donati in his "Roma vetus ac recens: utriusque aedificiis ad eruditam cognitionem espositis", for a graphic reconstruction of the area between Porta Flaminia and Campo Marzio. The drawing of the Augusto's Sundial clearly presents many useful and original details, particularly the special orientation of the obelisk and his pedestal, on the meridian line (Fig. 7). So, for our reconstruction of the Sundial, we have chosen the Buchner's model, because of its higher

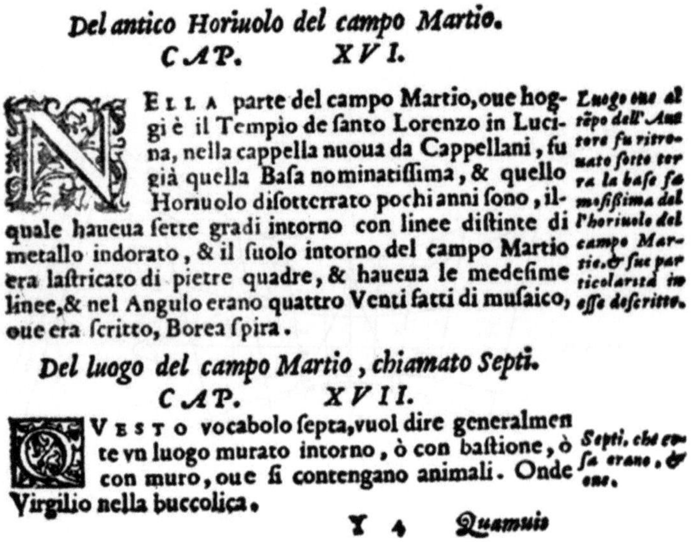
Del antico Horiuolo del campo Martio.

CAP. XVI.

NELLA parte del campo Martio, oue hoggi è il Tempio de ſanto Lorenzo in Lucina, nella cappella nuoua da Cappellani, fu già quella Baſa nominatiſſima, & quello Horiuolo diſotterrato pochi anni ſono, il-quale haueua ſette gradi intorno con linee diſtinte di metallo indorato, & il ſuolo intorno del campo Martio era laſtricato di pietre quadre, & haueua le medeſime linee, & nel Angulo erano quattro Venti fatti di muſaico, oue era ſcritto, Borea ſpira.

Luogo oue al tẽpo dell'Autore fu ritronato ſotto terra la baſe famoſiſſima dal l'horiuolo del campo Martio, & ſue particolarità in eſſo deſcritto.

Del luogo del campo Martio, chiamato Septi.

CAP. XVII.

QVESTO vocabolo ſepta, vuol dire generalmente vn luogo murato intorno, ò con baſtione, ò con muro, oue ſi contengano animali. Onde Virgilio nella buccolica.

Septi, che coſa erano, & one.

Y 4 *Quamuis*

Fig. 6 Description of the "Horiuolo" discovery at Campo Marzio, from L. III "De antiquities of Rome" by A. Fulvio, (courtesy of Vallicelliana Library)

relevance with the prevailing historical and literary sources. Due to the lack of time, other original and useful notes of history of astronomy, metrology and on calendarial's history, obtained in our work, are not here reported.

For the Meridiana Clementina of Francesco Bianchini (1662–1729), it has been realized an enhancing reality, to draw on the floor, along the bronze meridian line, the daily sun ellipse for any time. About information on the stony materials used for the zodiacal drawings inspired by the "Uranometria" of J. Bayer (1572–1625), along the 44.2 m of length of the meridian line we have used an original petrographic research, realized with the collaboration of Science Department of University of Roma Tre (Catamo and Lucarini 2002). Other information are available for the APP's users, particularly regarding the extraordinary accuracy of the astronomical measures, obtained by Francesco Bianchini, for specific calendarial research and for the studies of Earth's movements in the space. At present, the App STAR include two historical sites: Pantheon and St. Peter's Meridian. The other six sites are mithraic shrines: Barberini mithraeum, San Celmente mithraeum, Santo Stefano Rotondo mithraeum, Caracalla mithraeum, Santa Prisca mithraeum, Circus Maximus mithraeum.

It's worth mentioning, that the museographic works on mithraic sites in Rome (among those opened), have given original results: the variable orientation of the axes of the shrine places, in relation to the building's epoch; some enigmatic astronomical features in the frescoes of Barberini's Mithreum, possible extraordinary astronomical events, like planetary superconjunctions, to inspire the building of some mithraic cult sites.

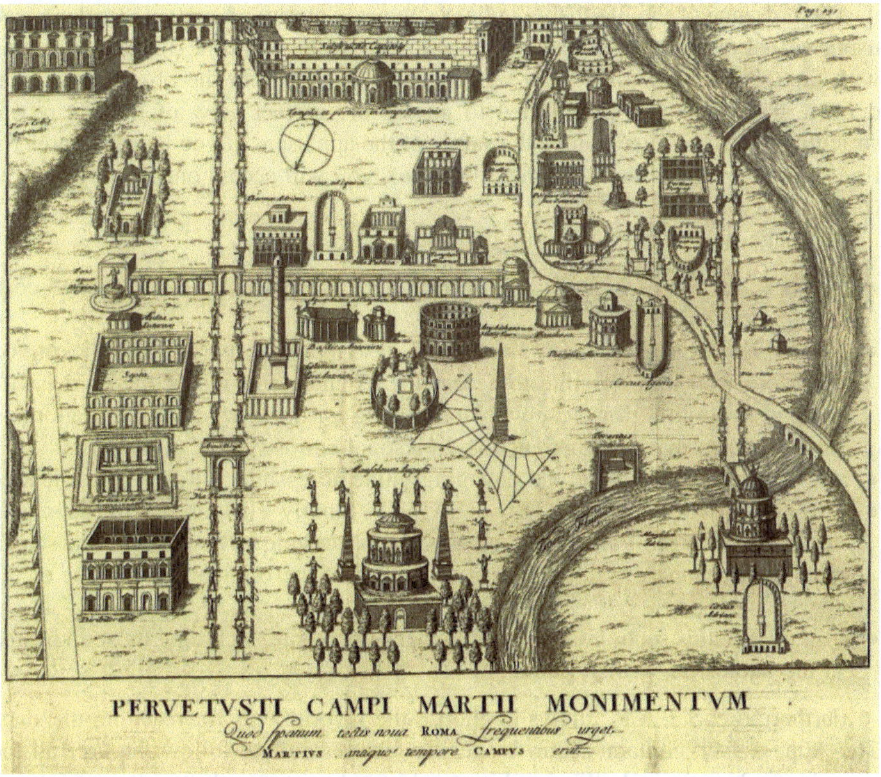

Fig. 7 Representation of "Horologium Augusti", the Mausoleum and the Ara Pacis, by A. Donati (op.cit.), in detail (Courtesy of Biblioteca Vallicelliana)

4 App's Use and Potentiality

To establish the eastern boundary of the Ekumen, Marinus and Ptolemy can rely only on data from travel reports. The Alexandrian mathematician mentions the astronomical method based on the observation of lunar eclipses, but states that he does not have a sufficient amount of data for a complete discussion. He then decides to take the information contained in the *Commentarii*, to process and correct them. From the Home menu you can visualize all STAR sites on the map of Rome. By clicking on each site, a popover window will appear showing its name. A set of intuitive icons below will guide the user to any information regarding the chosen site. A traffic-light color coding system is used to describe the availability of the site: (green) Open to the public, (yellow) Booking in advance required, (red) Not yet accessible, but you will be able to visit remotely thanks to the STAR multimedia content.

The presence of the 3D ICON indicates the availability of contents implemented using virtual or augmented reality. You can access the real-time 3D reconstruction with first-person view only once reached the site and from the observation areas defined. This area was established to ensure the correct view of the 3D reconstruction as well as to guarantee the safety of the app users. If you leave the area (or if you want to access the 3D content off-site), you will access only the flyby view (from above).

The page dedicates to each site contains:

- GENERAL INFO regarding the site such as booking details, ticket cost, opening/closing time.
- WRITTEN GUIDE that contains a brief description of the site. By clicking on the More icon it is possible to access the extended version of the guide, with more in depth info, for those who are looking for more detailed information, on the basis of previous experience and personal curiosity.
- AUDIOGUIDE a simple and engaging guide accessible with a single click.
- VIRTUAL OR AUGMENTED REALITY Simply clicking on the icon and aiming the device in the direction shown, it is possible to visualize on the screen of the smarphone or tablet the reconstruction of the original appearance of the site.
- GALLERY this section includes all images of the site, including those indicated in the audio and written guide.

Furthermore, S.T.A.R. allows to choose the favorite way for reaching the sites. The App, in fact, contain within it specific features able to follow the user all the time during his trip and allowing him to choose how to reach the sites of interest between WALKING, where the integrated S.T.A.R. Navigator suggest the quickest way to walk to the site, or using PUBLIC TRANSPORTS, where S.T.A.R. provide all necessary information (such as bus stop/tube ID, line name, average time and more) to guide to the destination.

Then, S.T.A.R. elaborate the quickest journey that best suits the user using as a starting and finishing point his current location.

In the section FOR MORE INFORMATION, S.T.A.R. includes:

- "FOR THE BOOKWORMS": list of the in depth resources organised by topic, site and more.
- "ROME IS STAR SHAPED": a city shaped like a Sun, symbol of power at the centre of the Mediterranean and the Ecumene in general, that would radiate civilisation. A fascination more than two millennia old, buried in small references in the classic texts, inspired by monuments and building rich in symbolisms, that like a STAR shed light on the discoveries and vision of scholars and researchers.
- "HOW DOES A SUNDIAL WORK?": the science behind this fascinating ancient technological device.
- "WHAT SOLAR TIME IS IT?": toolkit that allows to automatically calculate the solar time based on the time on your watch.

Acknowledgements The app STAR was realised thanks to the Programma Operativo Regione Lazio—European funds.

The authors would like to thank for the software developed the MobileSoft srl. (Walter Cavalcante and Laura Berellini), for the 3D reconstruction the Illusion network (Fabrizio Terranova and Gabriele Leonardi), for the translations the AlfaBeta—Language and Web Consultants and finally for the audioguides the Voice Professionals Italy (STAR is voiced by Maurizio Bruno and Olga Leo).

The authors would especially like to thank the University of Rome "La Sapienza", particularly Prof. Luigi Campanella and Prof. Giorgio Manzi, holders of the course of Museology during which were elaborated the museological contents used as a starting point for the development of STAR. In this regard, we would like to thank all the students of the course of Museology (Master Degree in Technology for the Conservation and Restoration of Cultural Heritage, University of Rome "La Sapienza").

Special thanks to the Direction of the Biblioteca Vallicelliana, for the aid and the kindness during the bibliographic research on rare books, fundamental to get rare iconographic materials.

References

Bandini, A.M.: De obelisco Caesaris Augusti, Ed. Roma (1751)

Catamo, M., Lucarini, C.: Il cielo in basilica, Ed. Agami (2002)

Donati, A.: Roma vetus ac recens: utriusque aedificiis ad eruditam cognitionem espositis, Roma: Editio secunda correctior (1648)

Fulvio, A.: De le antichità di Roma, Ed. Venezia (1588)

Heslin, P.: Augustus, Domitian and the so-called Horologium Augusti. J. Rom. Stud. 97, 1–20 (2007)

Lugli, P.M.: L'Agro Romano e l' "altera forma" di Roma antica, Ed. Gangemi (2006)

Malviani, B.: Le antichità di Roma, Ed. Roma (1632)

Plinio, G.S.: Storia naturale, Ed. Einaudi (1982)

Pomponio, L.: Excerpta a Pomponio dum inter ambulandum cuidam domino ultramontano reliquias ac ruinas urbis ostenderet, Ed. Valentini & Zucchetti (1953)

MIX
Papier aus verantwortungsvollen Quellen
Paper from responsible sources
FSC® C105338

If you have any concerns about our products,
you can contact us on
ProductSafety@springernature.com

In case Publisher is established outside the EU,
the EU authorized representative is:
Springer Nature Customer Service Center GmbH
Europaplatz 3, 69115 Heidelberg, Germany

Printed by Libri Plureos GmbH
in Hamburg, Germany